Vortex Publishing LLC.
4101 Tates Creek Centre Dr
Suite 150- PMB 286
Lexington, KY 40517

www.vortextheory.com

© Copyright 2019 Vortex Publishing

All rights reserved. No part of this book may be reproduced or transmitted in any form or by any means, electronic or mechanical, including photocopying, recording or by any information storage and retrievable system without the prior written permission by the Publisher. For permission requests, contact the publisher.

Printed in the United States of America

1 2 3 4 5 6 7 8 9 10

Library of Congress Control Number: 2019953543

ISBN   978-1-7332996-4-0
eISBN 978-1-7332996-5-7

Editor's note: All drawings in this book are original illustrations made by Dr. Moon. They are kept as they are to maintain the integrity of his work.

TABLE OF CONTENTS

Overview III
Author's note IV

PART I
THE PARTICLES OF THE NATURE & THE END OF THE UNIVERSE

Chapter 1: The Problem With the Construction of the Universe 1
Chapter 2: Assessment of the Quark Theory 3
Chapter 3: Quarks Are Higher Dimensional Holes in Space 5
Chapter 4: The Two Sides of Space 7
Chapter 5: Creation of the 1/3 and 2/3 Charges 9
Chapter 6: The Actual Charges on Quarks Are Not Percentages of ± 1! 12
Chapter 7: Creation of the ±1 Charge, the ±2 Charge and Spin 15
Chapter 8: The True Vision of Space 19
Chapter 9: Creation and Destruction of the Universe 24
Chapter 10: Creation of the UP and DOWN Quarks 29
Chapter 11: The Creation of the Strange and Charm Quarks 32
Chapter 12: Creation of the Top and Bottom Quarks 35
Chapter 13: "The Four Layers of Matter" 39
Chapter 14: The Weak Force 41
Chapter 15: Tunneling 44
Chapter 16: Neutrinos 47
Chapter 17: Tunneling, the Creation of a New Class of Particle 49
Chapter 18: Tunneling Particles 53
Chapter 19: The Asymmetric Parity of Neutrinos 55
Chapter 20: Charged Leptons: [the Electron, Muon, & Tau] 59
Chapter 21: Neutrino Oscillations 70

PART II
GREAT PARTICLE MYSTERIES OF THE UNIVERSE FINALLY EXPLAINED

Chapter 22: The Stability of the Proton; The Instability of Mesons 76
Chapter 23: The Explanation of the "Conservation" of Strangeness 79
Chapter 24: Gauge Bosons Are *NOT* Force Carriers Between Particles 80
Chapter 25: The Explanation of the Pauli Exclusion Principle 84
Chapter 26: Explanation of the CPT Theorem 88
Chapter 27: Motion of Photons and Particles Through Electric and Magnetic Fields 93
Chapter 28: Why the Electron Can Tunnel but Ordinary Matter Cannot 95
Chapter 29: Eliminating Popular Misconceptions 96
Chapter 30: Major Problems With Today's Popular Theories 98
Chapter 31: Why All "Particles" Possess the Same Amount of Charge 99
Chapter 32: The Explanation of the Strong Force 100
Chapter 33: The Alpha and Beta Particles 104
Chapter 34: Explanation of the Double Slit Interference Patterns Created by Electrons! 106
Chapter 35: How "Looking" at the Electron in the Two Slit Experiment Changes the Results!. 109
Chapter 36: Why the Electron Does Not Like to Be Confined 110

I

PART III
PARTICLE DECAYS AND COLLISIONS

- Chapter 37: Decay of the Neutron 111
- Chapter 38: New Particle? The Tunneling Pion! 115
- Chapter 39: Schematics of Higher Dimensional Particles 123
- Chapter 40: How Particle Collisions Create New Particles 124
- Chapter 41: The Explanation of How Quarks Change "Flavor" 127
- Chapter 42: Lepton Creation During the Decay of Positive and Negative Pions 132
- Chapter 43: Neutrino Creation During the Decay of the Positive Muon 135
- Chapter 44: Neutrino Creation During the Decay of the Negative Muon 138
- Chapter 45: The Decay of the Positive Tau 141
- Chapter 46: The Decay of the Negative Tau 143
- Chapter 47: The Decay of the "WOW" Lepton 146
- Chapter 48: The Collision Between a Proton and an Electron Anti-neutrino [And: The Proton, Muon Anti-neutrino Collision] 147
- Chapter 49: The Collision Between a Neutron and an Electron Neutrino; And the Decay of The Neutron and the Creation of the Anti-neutrino 150
- Chapter 50: The Decay of the Neutron and the Creation of the Anti-neutrino 152
- Chapter 51: The Stability of the Proton; the Instability of Mesons 154

PART IV
SIX CONSERVATION LAWS OF NATURE FINALLY EXPLAINED

- Chapter 52: Explanation of the Conservation Law of Lepton Number 156
- Chapter 53: The Explanation of the Law of the Conservation of Baryons 174
- Chapter 54: The Explanation of the Conservation of Momentum 176
- Chapter 55: The Explanation of the Conservation of Angular Momentum 177
- Chapter 56: The Explanation of the Conservation of Charge 180
- Chapter 57: The Explanation of the Conservation of Mass and Energy 181

PART V
QUANTUM ENTANGLEMENT

- Chapter 58: The Explanation of Quantum Entanglement 182

OVERVIEW

This is the third book in the six part series called The End of the Concept of Time.

In this book, the creation of the particles of nature are explained. Also, other concepts that are needed to explain numerous aspects of the construction of the universe are added. In fact, so many are presented in this work that they are too numerous to mention all of them here. A brief glance at the index will do a better job of describing what they are rather than listing all of them here.

None-the-less, several have to be mentioned here. They have to because they are explanations for phenomenon no-one or no theory has ever explained before!!! Some of them are: what the ±1/3 & ±2/3 charges of quarks really are; why a -1/3 charged particle can change into a +2/3 particle and vice versa; *the true explanation of Quark Confinement*: the true explanation of "tunneling", and the creation of a new class of tunneling particles. Next comes the Asymmetric Parity of Neutrinos; the explanation of the Pauli Exclusion Principle; the explanation of the CPT Theorem; the unification of Quantum Mechanics and Newtonian Physics via NUCLEAR GRAVITY; and the explanation of Nuclear Gravity! The explanation of how "Looking" at the electron in the two slit experiment changes the results; the discovery of a new particle in nature the Tunneling Pion! And a second new particle - the fourth lepton – the "WOW" Lepton [a massive charged particle much larger than the tau]! Then there is the shocking discovery that Gauge Bosons are *NOT* force carriers between particles; the Explanation of the Conservation Law of Lepton Number! And perhaps one of the most shocking discoveries of all: the Creation of Yukawa's Virtual Particle that is found to be the true explanation of the strong force!

Never has any book possessed so many revolutionary discoveries as those you are about to encounter!

AUTHOR'S NOTE:

This first ever book uses pictures to explain what was previously one of the most difficult subjects in all of physics. But no more! This revolutionary breakthrough in science now makes it possible for High School children [14-18 years] to understand what the greatest scientists of the 20th Century were unable to comprehend. Although some minor math is necessary, it only takes an eighth grade education to understand it.

Also, even though the subatomic particles of nature are the subject of this book, and are talked about in the first chapter, making it seem as if a special education is necessary to understand them. <u>This is not so</u>. Later on, all of the terms in the first chapter are easily explained so that everyone can understand them. And it is very important that everyone does understand them: because these infinitesimal particles are the key to understanding our ultimate fate! So do not let the misconception that you have to be a rocket scientist cheat you from learning the most fantastic discovery ever made about yourself and your universe.

To all you good friends and to all you good friends I will never meet, welcome to the third book in this series, *The End of the Concept of Time PART 3: The Quark Theory.*

Like the first two books, this one presents revolutionary knowledge about what can only be called the "Midnight World of Microscopic Space." Using the principles of what has come to be known as the Vortex Theory of Atomic Particles, the discipline of physics that has come to be called "Quark Theory" is revolutionized. Nothing will ever be the same again.

In honor of two of Russia's greatest scientists: Dr., Prof. Konstantin Gridnev; and Dr., Prof. Victor V. Vasiliev, I have given the vortices the name "Konsiliev Vortices".

Sincerely,

Russell Moon

PART I
THE TRUE VISION OF THE UNIVERSE

Chapter 1
The Problem With the Construction of the Universe

> The mystery of the construction of the universe appeared to be solved with the ideas of the "Big Bang" and what is called the "Singularity." But on second glance, serious problems are revealed in these solutions.

Many years ago, when I first set out upon the quest to discover the secrets of the construction of the universe, I, like many others went to College. There I thought I would find the answers. Instead, what I did discover was that the professors teaching the courses I took were themselves lacking answers just as I was. They knew many theories and had extensive knowledge of mathematics and other branches of science, but when it came to the construction of the universe there were just too many unanswered questions. For example…

We have all heard of the Big Bang, the giant explosion that began the universe and set matter into motion; and of course, the "thing" that exploded: the "Singularity," the micro-sized little dot that contained all of the matter of the universe within it. However, a little research into the history of the Big Bang now reveals a big problem! This theory was proposed almost a hundred years ago before the hundred billion plus galaxies were discovered that are now known to populate the universe. So how can all the matter in a hundred billion plus galaxies be compressed into a single dot no bigger than a proton?

The individual who came up with this theory was not Edwin Hubble as the history books proclaim; instead, it was Georges Henri Joseph Édouard Lemaître: a Belgian Catholic Priest, astronomer and professor of physics at the Catholic University of Leuven. Lemaître published his theory in 1925, two years before Hubble published his article in 1927. In 1925, Lemaître proposed the Big Bang, the Singularity [also called the "Cosmic Egg"], and the expansion of the universe! But if he knew there were a hundred billion galaxies, would he still try to theorize that all of the matter – in all of these galaxies – was compressed into one single dot?

He would also be faced with the problem of Black Holes. When matter is compressed into a Black Hole, its volume grows larger not smaller. If Lemaître was right about the Singularity, Black Holes would grow smaller and smaller instead of larger and larger.

And how about the Big Bang? This theory is unequivocally true say scientists. The discovery of the "Background Radiation" reveals that the Big Bang indeed happened. Also, the "red shift" of the light coming from distant galaxies reveals that the universe is expanding outward as a result of the Big Bang. All true…until you ask this one little question, "Where is the Point of Origin of the Big Bang?" "Or more specifically, where is the point of origin where the Singularity exploded and began its expansion?"

As was mentioned in Book 2 of this series, in 1054 AD, the people of the earth witnessed a gigantic supernova. Although they did not know what they were seeing, the Chinese wrote about a new star that suddenly appeared in the sky and was so bright that it could be seen in the middle of the day. In the southwest, Native Americans drew stars on rocks in awe of this one of a kind event. Today, in our telescopes, we can witness the remnants of that explosion in the Crab Nebula.

By looking at the Crab Nebula and measuring the Doppler shift of its light, we can tell by the velocity of the particles streaming out into the universe that the nebula is the result of a gigantic supernova explosion that occurred approximately 1000 years ago [approximately equaling the year 1054 AD]. We can also use the same technique to trace the coordinates of the particles streaming out into the universe from the Crab Nebula to determine the point of origin of this supernova. Hence, we can see approximately where the blast occurred.

So, how come we cannot do this with the Galaxies of the universe? How come we cannot use this same technique to determine the point of origin of the Galaxies created after the Big Bang?

Unfortunately, this same scientific method fails when we try to apply it to the motion of the Galaxies out of which the physical universe is constructed. Even more amazing, we find it hard to believe that when we look for the origin of the Big Bang, we discover *that **every point in the physical universe** appears to be the point of origin we are looking for! How can this be?* [This observation will become extremely important later on when we discover the true shape of the universe.] Because if indeed the Singularity existed, the point of its explosion outward would, like the Crab Nebula, have one single point of origin and not an infinite number of points!

It was this failure to explain the Point of Origin that made me realize something is wrong!

So how can this problem be resolved?

At the time I was introduced to this problem, I was just a freshman student without the knowledge necessary to offer a solution. However, all that is about to change.

A strange set of circumstances forced me to re-investigate what has come to be called the Quark Theory. Amazingly, found in the knowledge explaining the tiniest of all the pieces of the universe - is also found the explanation of the most massive aspects of it!

This book is a result of this investigation, and the discoveries presented here allow us to finally explain the creation of the universe. And it is indeed a shocker, unlike anything anyone has ever imagined before!

Throughout this book, we will present the knowledge allowing all of us to discover how this universe we live and exist in was created. And more important – our destiny!

Chapter 2
Assessment of the Quark Theory

> What looks good at first is nevertheless found to also possess as many mysteries as it is touted to explain. [All of which will be explained in this book.]

When we first look at the Quark Theory, it appears to be one of the most successful theories ever. Using the fundamental principles such as mass, ½ spin, ±1/3 and ±2/3 charges, and conservation laws, the Quark Theory is used to classify all of the hundreds of particles known to science as hadrons (particles made up of quarks such as the proton).

The Quark Theory has also allowed scientists to discover that hadrons are split into two families: the Mesons, containing two quarks; and the Barons, containing three quarks. Even more important, the use of the Quark Theory has allowed theorists to predict the existence of new particles and then look for them in the photographs taken during particle collisions in accelerator experiments. These efforts have not gone unrewarded.

Many Nobel prizes have been awarded for these discoveries. And with governments spending hundreds of millions of dollars on the experiments used to make these discoveries, many physicists have prospered.

Theorists make a good living teaching the Quark Theory and using it to study particle collisions. The science departments in universities have prospered by offering many courses relating to the Quark Theory. Adding up all of the accolades and rewards, in appearance, the Quark Theory appears to have been one of the most successful theories used in making important scientific discoveries, increasing man's knowledge of the universe, and creating an economic boom for the scientific community. However, on closer inspection, there is a serious problem with the Quark Theory.

The Quark Theory cannot explain the fundamental principles upon which it is based: the Quark Theory is based upon the conservation of charge [explained in Book 2]; the conservation of baryons [that will be explained in this book]; and the conservation of leptons [also to be explained in this book]. And yet, the Quark Theory cannot explain why these three conservation laws exist [we will]! Quarks are also classified according to three important physical characteristics – their spin, charge, and mass. However, the Quark Theory is at a loss to explain why quarks possess spin, charge, and mass. For example:

The Quark Theory states quarks possess what is called "spin up" or "spin down" [clockwise or counterclockwise rotations], and yet the Quark Theory cannot explain why quarks possess spin up or spin down [We can, it was done in Book 2]! Quarks are also said to have -1/3 or +2/3 charges, but again, the Quark Theory cannot explain why quarks possess these charges, *and even more important – how do quarks generate these charges (?)* [We will]. And then there is mass.

From our practical everyday point of view, the most important characteristic of matter is mass. We absolutely have to know what the mass of matter is in order to perform all of our engineering and scientific calculations. However, if quarks are the fundamental particles responsible for the mass of all the matter we use in our calculations, how come science does not know what creates mass [in Book 2, mass was explained; and also revealed that the so called scientific explanation for mass – the "Higgs Boson Particle" is **WRONG**]!

But this is only the start of the problems with the Quark Theory, there are many more. One of the most important ones is how a quark can change *"Flavor"* [change from one type of a quark into another type of quark]. It is proposed in the Quark Theory that quarks change from one kind of a quark to another type of quark, yet no explanation is given, or even speculated upon [we will explain it in this book]. Even more bizarre is the shocking fact that quarks with –1/3 charges can change into quarks with +2/3 charges and vice versa, but nobody can explain the mechanism via which this transformation occurs [we will].

Anti-quarks also create another set of problems. Why do anti-quarks exist? [Why does any anti-matter exist?] Why is an anti-matter particle created when a regular particle of matter is created [we will explain this]? What mechanism in the subatomic world is causing this phenomenon called *Associated Production* [we will do it]?

And what about leptons? Why do Pions and Kaons [that have two quarks] decay into leptons [such as the electron] that don't have any quarks [again, we will do it]?

When it comes to leptons [electrons, muons, tau particles and neutrinos], the Quark Theory fails badly again. The Quark Theory has no explanation for the construction of leptons. It cannot explain why the electron, muon, and tau particles that have no quarks, nevertheless still possess the same charges as the hadrons that *are* constructed out of quarks. The creation of electrons, muons, and taus are also a mystery. Why only these three particles? How come there are no other intermediary particles with masses in between the electron and muon or the muon and tau [we will explain everything in this paragraph]? [Note: neutrinos were also explained in Book 2.]

The laws of the relationships between leptons are also mysteries. Although it is known what neutrino particles are created with electron, muon, or tau particles, how and why they are created is yet another mystery; and it is also a mystery as to how and why neutrinos change protons into neutrons, and neutrons into protons [we will do it in this book].

The contradictory nature of neutrinos is also a mystery. How come neutrinos travel at the speed of light and yet interact with matter as if they had mass? Equally intriguing is the asymmetric parity of neutrinos. How come all neutrinos only spin in one direction while all anti-neutrinos only spin in the opposite direction [we will explain this too]?

Then there is the science of Quantum Chromodynamics. It is a beautiful theory, and yet, the science of Quantum Chromodynamics creates as many problems as it explains. Perhaps the most important is this one: Quantum Chromodynamics uses its three color charges to satisfy the Pauli Exclusion Principle, *but has no explanation for how these charges are created* [again done in this book].

And then there is the Pauli Exclusion Principle itself. What is its explanation? How come the billion dollar particle accelerator machines have not even helped to explain one of the most basic fundamental principles of matter used by every student of chemistry in the world [again, we will do it]!

Although there are many more problems, the ones listed demonstrate that the science of sub-atomic particles is lacking and possesses as many mysteries as it explains. However, in this third part of the Vortex Theory, this situation will be rectified. All of the mysteries mentioned above - along with many others - will be easily explained.

Chapter 3
Quarks Are Higher Dimensional Holes in Space

> Quarks are holes in higher dimensional space. Shockingly, they are holes within holes! Also, this relationship explains why quarks cannot exist outside of other "particles" such as the Proton: explaining the phenomenon called "Quark Confinement." [This idea that quarks are higher dimensional holes in space, reveals that the universe is constructed out of a number of dimensions; and gives us one of the first clues to the construction of the universe.]

As mentioned earlier in Book 2, the construction of space is the key to understanding how everything else in the universe is constructed. Using this line of reasoning and beginning with the relationship between a surface and the volume it encloses, the explanation for the construction of quarks suddenly becomes quite easy.

Just as protons and electrons are three dimensional holes existing upon the surface of fourth dimensional space, *quarks are 4th, 5th, and 6th dimensional holes existing upon the surfaces of the higher dimensional space trapped within the interior of these three dimensional holes*. To understand how this relationship works, it is extremely important to realize that each lower dimension is the surface of the next higher dimension.

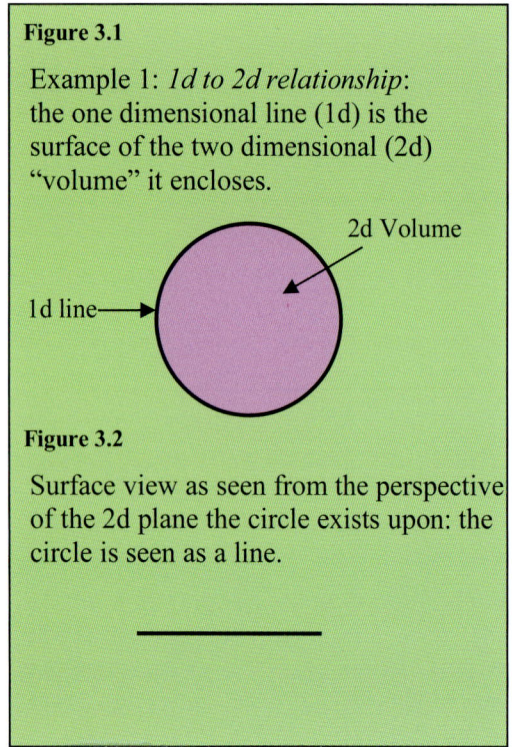

Figure 3.1

Example 1: *1d to 2d relationship*: the one dimensional line (1d) is the surface of the two dimensional (2d) "volume" it encloses.

Figure 3.2

Surface view as seen from the perspective of the 2d plane the circle exists upon: the circle is seen as a line.

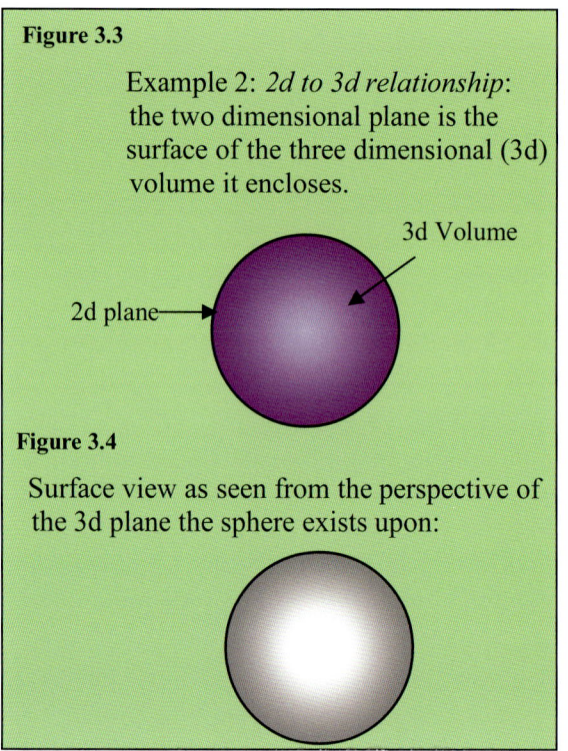

Figure 3.3

Example 2: *2d to 3d relationship*: the two dimensional plane is the surface of the three dimensional (3d) volume it encloses.

Figure 3.4

Surface view as seen from the perspective of the 3d plane the sphere exists upon:

Although these surface views seem meaningless now, later we will refer to them when we talk about the charges on quarks and suddenly we will see how important these two views are. But first, we must talk about holes and the surfaces they exist upon.

Because a hole in a surface is an entrance into the next higher dimension, just as a two dimensional hole is the surface entrance into three dimensional space, a three dimensional hole is the three dimensional surface entrance into fourth dimensional space; a fourth dimensional hole is the surface entrance into fifth dimensional space; while a fifth dimensional hole is the surface entrance into sixth dimensional space, and so on. When higher dimensional holes exist within lower dimensional holes, this effect can be called "*Layering.*"

The creation of this layering effect can be seen during the collisions between "particles." For example, when two three dimensional holes are traveling towards each other at high velocities, their momentum is high because their holes are greatly distorted into "pear shapes" as was seen in Book 2. The great distortion upon their surfaces gives them the ability to greatly distort space if they collide with each other. And if they do happen to collide, the tremendous warping created within three dimensional space can extend into fourth dimensional space warping it too. When this situation occurs, the collision can open holes in the fourth dimensional surface creating an entrance into fifth dimensional space.

This fourth dimensional hole created upon the surface of fifth dimensional space is a quark: an *up* or *down* quark. If the collision is even greater, the warp can extend onto the fifth dimensional surface, tearing it and creating a hole upon the fifth dimensional surface extending into sixth dimensional space. This fifth dimensional hole creates another "layer" of quarks: the "*strange*" and "*charm*" quarks. And finally, if the warping is great enough, it can extend all the way into the sixth dimensional surface creating an entrance into seventh dimensional space. This final layer of holes creates the "*top*" and "*bottom*" quarks.

This idea is not speculation. It is based upon the decay of more massive particles into less massive particles. This will be discussed later.

The creation of higher dimensional holes "sheathed" within lower dimensional holes finally allows us to be able to explain the mystery of "quark confinement."

EXPLANATION OF QUARK CONFINEMENT:

Just as a fish is trapped within the water of a fishbowl, quarks are trapped within the volume of higher dimensional space sheathed within a lower dimensional hole. For example, what are called Up and Down quarks are fourth dimensional holes trapped within the fourth dimensional space sheathed within three dimensional holes; Strange and Charm quarks are fifth dimensional holes trapped within the fifth dimensional space sheathed within fourth dimensional holes; and Top and Bottom quarks are sixth dimensional holes trapped with the sixth dimensional space sheathed within fifth dimensional holes.

Using the above relationships, it is now clear why a quark cannot exist alone as a single "particle"-unique unto itself. Since our universe is really the three dimensional surface of a fourth dimensional volume of space, any hole penetrating into the fourth dimensional volume has to first begin upon the three dimensional surface. Because each higher dimensional hole has to be sheathed within its lower dimensional surface, as progressively higher dimensional holes are created, they can only be created after a previous hole was first created upon their lower dimensional surface. This creates a hierarchy of holes forcing each higher dimensional hole to exist within a lower dimensional hole.

This idea is no speculation but is based upon the decay of massive particles. It is revealed in accelerators that for Leptons, the Tau decays into the less massive Muon, that in turn decays into the less massive electron. [The decays of the Top, Bottom, Charm, Strange, Up and Down will be discussed later.]

Chapter 4
The Two Sides of Space

> Just as a two dimensional surface has two sides, our three dimensional space has two sides; the fourth, fifth, sixth, and seventh dimensional space all have two sides also. These two sides of each dimension create a number of extraordinary phenomena. [This idea of two sides of space and two volumes of space gives us our next clue as to how the universe is constructed.]

Just as important as the number of dimensions that exist are the number of sides of space that exist too.

In Book 1, , the way to visualize the entrance into fourth dimensional space was introduced. It was noted that each higher dimension is at right angles to all of the lower dimensions simultaneously. For those who have not studied this relationship it can be explained as follows:

Figure 4.1

The line represents the first dimension. The second dimension is at right angles to the first dimension;

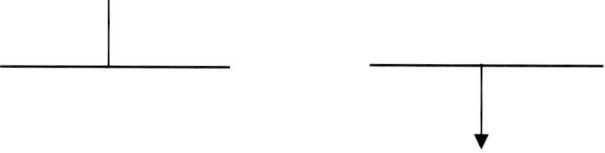

Figure 4.2

The third dimension is at right angles to the first and second simultaneously;

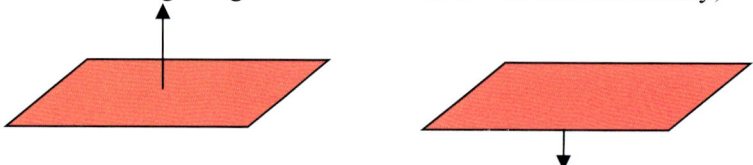

Figure 4.3

The fourth dimension is at right angles to the 3rd, 2nd, and the 1st simultaneously;
[This figure is impossible to draw. However, it can be represented by a series of arrows within a three dimensional sphere pointing towards or away from its geometric center.]

Figure 4.4

The fifth dimension is at right angles to the 4th, 3rd, 2nd, and the 1st simultaneously; the sixth is at right angles to the 5th, 4th, 3rd, 2nd, and the 1st simultaneously; and the seventh is at right angles to the 6th, 5th, 4th, 3rd, 2nd, and the 1st simultaneously.
[These figures are also impossible to draw.]

Looking at the above relationships, it is also easy to realize that each higher dimension possesses two different directions [one direction points towards and the other away from the higher dimension]. For the three lower dimensions these directions have already been defined: forwards and backwards, left and right, and up and down; and in Chapter 4 *The Vortex Theory of Atomic Particles,* it was explained how the fourth dimension possesses two new directions that can be called "within", and "without". The importance of these opposite directions reveals that each higher dimension of space also possesses *TWO SIDES*!

Just as a two dimensional surface has two sides, our three dimensional space has two sides and the fourth, fifth, sixth, and seventh dimensional space all have two sides also.

The way to envision the two sides of any dimension is to imagine a hollow sphere. While viewing the sphere in your mind it is easy to see that there are two sides to its surface. One side faces the center of the sphere while the other side points away from the surface:

Figure 4.5

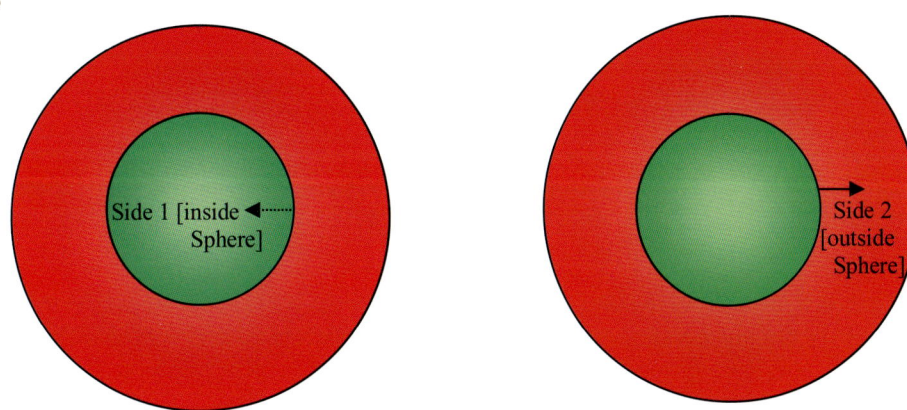

The side facing the center [the inside] can be called *Side 1*, while the opposite side [the outside] can be called *Side 2*. In the above figures, note how one arrow points to a three dimensional volume trapped within the sphere; while the other points to a three dimensional volume outside of the sphere. Using these two sides of space, we are almost ready to explain quarks; however, before we do, we must first introduce what can only be called, "the two volumes of space."

THE TWO VOLUMES OF SPACE

Many years ago, my preliminary investigations of the construction of nucleons allowed me to hypothesize that we lived upon the expanding three dimensional surface of a volume of higher dimensional space. Because I did not know what this volume was expanding into, until other evidence presented itself, it was only logical to conclude that it was expanding into nothing at all - a void: but this was a mistake.

After the study of quarks began, I realized that the two different types of charges - the 1/3 and 2/3 charges - could be explained if two volumes of space were present instead of just one. These two volumes would be responsible for creating a mutually shared surface: just as the Earth's atmosphere and oceans share a mutual surface, a mutually shared surface would also exist between the two volumes of space. And equally important, if one volume was expanding into the other, then the surface of one volume would be in expansion and the surface of the other would be in contraction [creating two different Elastic Modulus' for each space].

So how can the mutually shared surface of the two volumes of expanding and contracting space create the 1/3 and 2/3 charges of quarks? The answer involves a beautiful trick of nature.

Chapter 5
Creation of the 1/3 and 2/3 Charges

> It is amazing to contemplate, but the creation of the "1/3 & 2/3 charges of nature is one of the key discoveries [#3] in explaining how the universe was created. Even more incredible, these charges are not 1/3 & 2/3 charges! Instead, one charge is actually twice the other!

In the 1960s and 70s, the search for and discovery of quarks was the major interest of physics. The discovery of the six types of quarks and their 1/3 & 2/3 charges was considered a major triumph of science. However, nobody has ever explained why quarks possess these particular charges? [What mechanism within quarks generates these charges?]

The scientists of this era merely ASSUMED that these type of charges had to exist to be able to add up to the value of the +1 fundamental charge of nature. For example, inside the proton, because its charge is +1, the three quarks within it have to add up to +1: [2/3 + 2/3 + (– 1/3) = +1]; or within a particle called a pion with a charge of +1: [+1/3 + 2/3 = +1].

However, the fly in the ointment occurs when it is realized that the tiny electron containing no quarks has a charge of – 1, and its anti-particle the positron has a charge of +1. So, what is going on here?

Well, the problem is not with nature, the problem is with the scientists who are trying to interpret nature. What they did not realize is that a +2/3 charge can also be interpreted to be a charge that is twice the +1/3 charge! Or, simply stated, one charge is twice the other! So how can something like this happen?

It can happen if there are two different volumes of space with one possessing twice the elasticity of the other. In other words, if the density of one volume of space is twice the density of the other, such a situation makes the volume of one vortex twice the volume of the other, making its surface area twice the area of the other, making the charge double the value of the other.

Such a condition in the universe could arise if there are two volumes of higher dimensional space, one in contraction, one in expansion with three dimensional space [the surface of both] trapped in the middle between them.

So how could such an unusual relationship between the two different elasticity's of the two volumes of space be created?

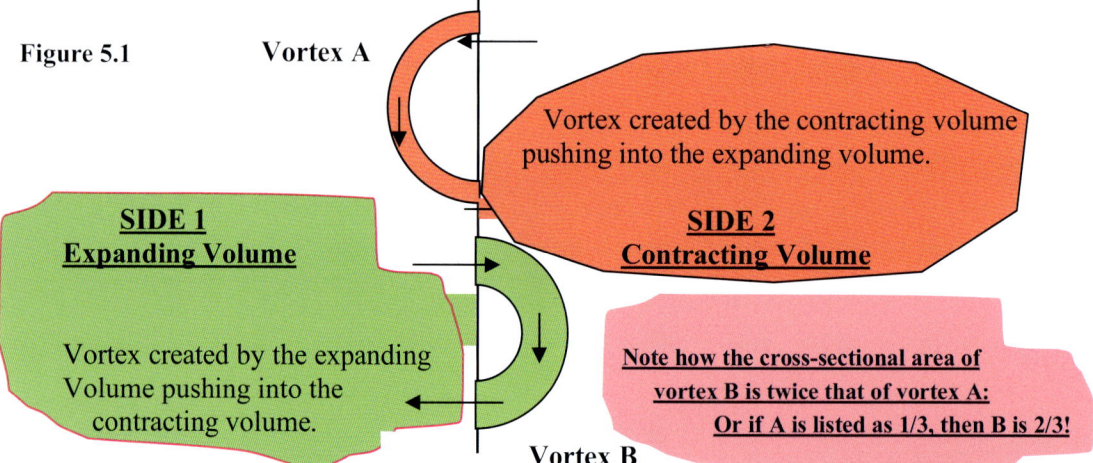

Figure 5.1

So how could such an unusual relationship between the different elasticity's of the two volumes of space be created? It can be explained using the analogy of two spheres: one in contraction, one in expansion:

This expansion and contraction can be more easily understood by using the analogy of two spheres – one inside of the other – and sharing the same volume. As the volume of the exterior sphere flows into the interior sphere, the volume of the interior sphere increases and the volume of the exterior sphere decreases. Because what flows out of one volume flows into the other, the rate of expansion is a *one to one ratio*: [as one cubic meter of space is added to the interior of the sphere, one cubic meter is subtracted from the volume of the exterior sphere]. Note in the figures below, that as the interior volume increases, the exterior volume decreases.

Figure 5.2

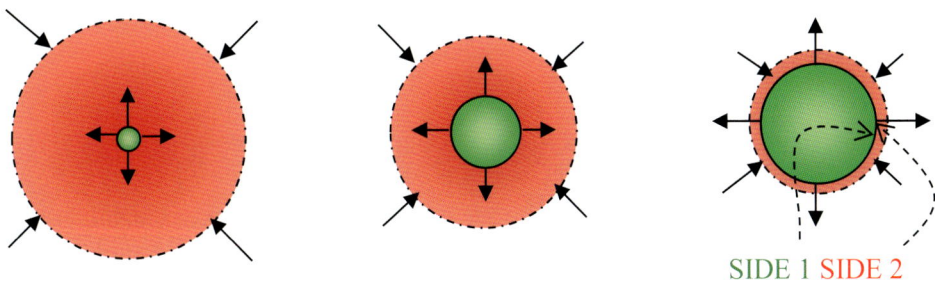

SIDE 1 SIDE 2

Although the flow from the contracting volume into the expanding volume is really a one to one ratio, from the relative perspective of either volume, a different relationship appears. From the relative perspective of either volume, the difference is actually **a *two to one ratio*!** [If the initial interior volume is 10 m^3 and the initial exterior volume is also 10 m^3, as space flows into the interior volume increasing its value to 11 m^3, the exterior volume of space decreases by 1 m^3/sec, decreasing its volume to 9 m^3.] Consequently, because the ratio is two to one, expressed in terms of pressure, it now becomes easier to push outward from the expanding volume [now designated as SIDE 1] into the decreasing volume [now designated as SIDE 2], than it is to push inward from the decreasing volume of SIDE 2 into the expanding volume of SIDE 1. This ability makes it appear as if the elasticity of each volume of space is different.

Transferring this concept to our physical universe and the two volumes of space, this 2 to 1 relationship affects every square meter of the mutually shared surface simultaneously: affecting the elasticity of the two volumes of space everywhere simultaneously.

[Note: this is not speculation. If the difference was not uniform, the flow of the vortices within atoms would not be the same; they would then release photons of differing velocities, (see: The Vortex Theory of Atomic Particles), and we would witness different spectral lines for the same elements in different Galaxies.]

In Figure 5.4, the effects created by the two different elasticity's is illustrated: if a vortex of flowing space pushes into the expanding sphere from SIDE 2 creating Vortex A with a flowing volume of 1 m^3/sec, then, Vortex B, Figure 5.3 created by pushing outward from SIDE 1 into the contracting sphere SIDE 2 will have a flowing volume of 2 m^3/sec - twice that of Vortex A.

| Figure 5.3 | Figure 5.4 |

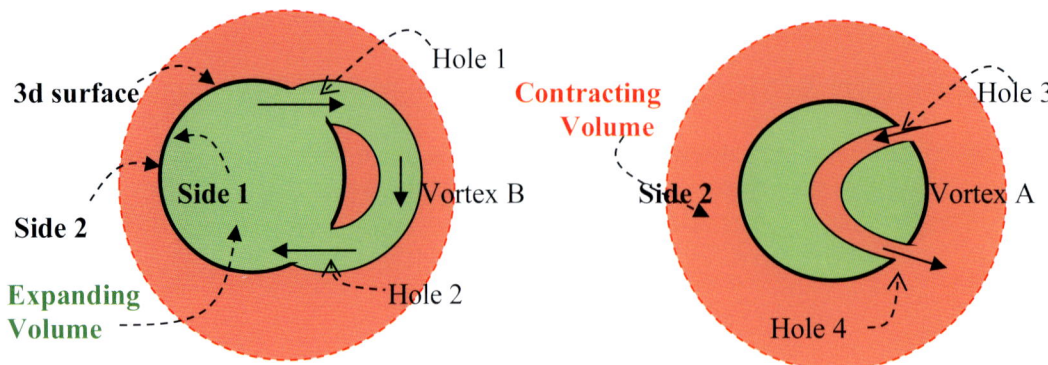

Because space is flowing into Holes #1, and #3, they are considered positively charged; while Holes #2 and #4 that have space flowing out of them are considered negatively charged.

[Although the velocity of space turning inside out is responsible for creating the speed of the flowing space in the vortices [the speed of light], this phenomenon will be discussed later, due to the fact that not only does it appear to be accelerating, but also, due to the incredible possibility that the speed of light only appears fast from our limited perspective!]

Because quarks are fourth dimensional holes trapped within three dimensional holes, and since it will be shown later how the charges on quarks are functions of the space flowing into the three dimensional holes [and not the other way around as is presently believed], then, when two or more quarks are trapped within a three dimensional hole whose three dimensional charge is equal to +1, the flowing volume is divided between them into values that only *appear* to be fractions: [+1/3 +2/3 = +1 (the charge on a positive pion); or, – 1/3 +2/3 +2/3 = +1 (the charge on a proton)].

And even though the charges appear to be divided up into thirds, it is now easy to see the absolute value of the 2/3 charge is really twice *the absolute value of the 1/3 charge.*

In our search for an explanation for the creation of the universe, it does not seem logical to assume that two different types of space exist with one possessing a greater elasticity than the other. But it does seem logical to theorize that the different dynamics created by the contraction of one volume and the expansion of the other could have created this unique circumstance. The following explanation is one theoretical possibility:

If the three dimensional surface upon which we live was created by a region of higher dimensional space turning inside-out, then the three dimensional space of our physical universe is really a line of demarcation between two volumes of higher dimensional space: one in expansion, the other in contraction.

Chapter 6
The Actual Charges on Quarks Are Not Percentages of ± 1!

> Before we can use the knowledge we have just found in the first five chapters of this book to explain the creation of the universe, a great fallacy regarding science's assignment of charges on quarks must be dispelled. And it is a shocker, because everything we have just described is only true from *our perspective!* When viewed from the ultimate perspective of the universe, none of what we have just seen is true! But how can this be?

To begin to grasp the enormous significance of the above statement we must first realize that the surface area of a higher dimensional hole *is much larger* than the surface area of a lower dimensional hole. It must also be mentioned that fourth dimensional holes possess four dimensions of flowing space. Consequently, the volume into and out of a higher dimensional hole is greater than the volume flowing into and out of the three dimensional holes they are contained within! Unfortunately, because our electronic instruments are constructed out of three dimensional holes, they can only measure the phenomena that affect three dimensional holes. And the only thing that affects three dimensional holes is three dimensional space.

These three dimensional instruments are incapable of measuring disturbances in fourth dimensional space that do not affect three dimensional space; they are only capable of measuring a disturbance taking place in the fourth dimension that directly affects its three dimensional surface. Consequently, from our point of view, the charges of the fourth dimensional holes only appear to possess three dimensional charges. Because this observation is one of the most profound concepts we will ever encounter in the world of quarks, the following two dimensional to three dimensional analogy of the "Sphere Within the Plane" is presented.

THE SPHERE WITHIN THE PLANE

To begin to understand why we only see part of the charge possessed by quarks, observe the following two illustrations. Note how a three dimensional sphere has been placed within a two dimensional hole existing upon a two dimensional plane. Note too, how each illustration represents a different perspective of the same two objects: the plane and the sphere.

Figure 6.1

Our three dimensional view:

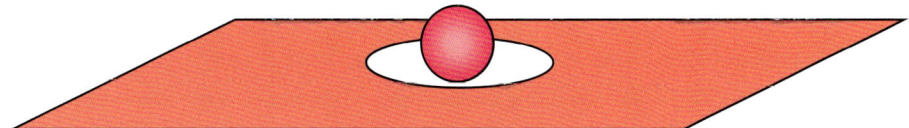

Figure 6.2

The side view of the 3d sphere, the plane, and the 2d hole. [Note how we cannot see the 2d hole.]

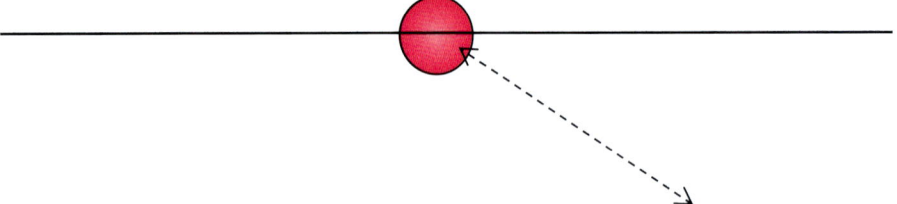

In the second figure, notice how the two dimensional plane can only intersect and or touch the three dimensional hole along one single plane. Realizing the sphere is actually constructed out of an infinite number of 2d planes allows us to realize that the 2d hole can only interact with the 3d sphere along *one and only one cross-sectional plane of the sphere*!

Figure 6.3

The dotted line represents the only place on the sphere that the 2d plane can interact with it.

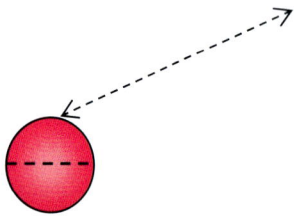

Now if the three dimensional sphere was in fact a 3d hole that 3d space was flowing into from all directions at once, from the ultimate perspective of the universe, the amount of the space flowing into the hole is a function of its square area [meters squared]. But the perspective from the plane is totally and completely different. From the perspective of the plane, interactions can only take place along the dotted line of the sphere.

This presents a serious problem because if an imaginary two dimensional creature living on the plane wanted to understand the construction of the object within the hole, whatever experiment he performs can only take place *along the dotted line*. If he tries to determine its charge by hurling other smaller 2d holes at it, they can only bend the surface of the 2d hole outward enough to touch the dotted line and react with the space flowing into it or out of it along the **dotted** line only. Hence, he can only perceive the space flowing into the 3d hole along the single dotted line drawn on the sphere. And this creates a problem.

This makes it appear as if the 3d object possesses the same [two dimensional] type of charge as the two dimensional hole it resides within. And suddenly we begin to realize that even though the imaginary 2d **creature's measurement of the charge** is absolutely wrong, from his point of view, he appears to be absolutely right. And this is the *exact same situation we are experiencing* living upon the 3d surface of 4d space.

From our point of view, quarks appear to possess 1/3 and 2/3 charges. But in reality, quarks possess fourth dimensional charges that are completely unlike the three dimensional charges we experience within our three dimensional universe.

Figure 6.4

Our three dimensional view, we see two quarks in a Pion:

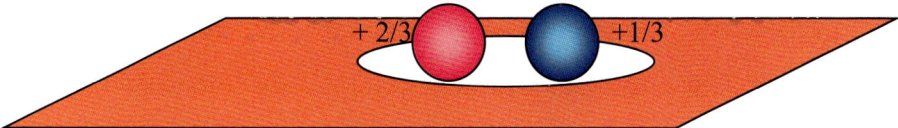

Figure 6.5

The side view of the 3d sphere, the plane, and the 2d hole. [Note how we cannot see the 2d hole.]

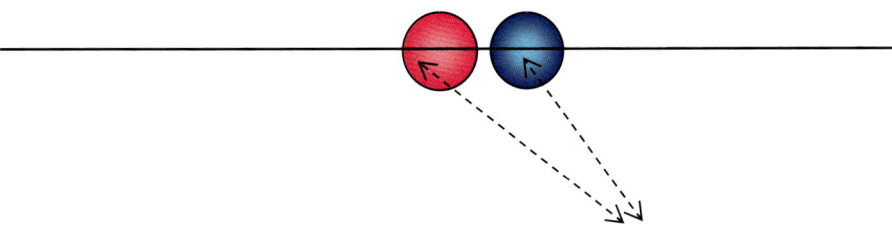

In the second figure, notice how the two dimensional plane can only intersect and or touch the 2 three dimensional holes along one single plane. Realizing the spheres are actually con-structed out of an infinite number of 2d planes allows us to realize that again, the 2d hole can only interact with the 3d spheres along *one and only one cross-sectional plane of the sphere*!

Figure 6.6

The dotted lines represents the only places on the spheres that the 2d plane can interact with them.

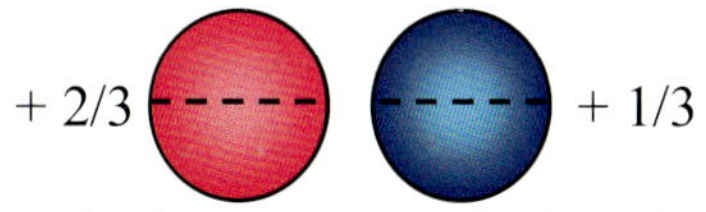

Unfortunately, we can only calculate mathematically what these fourth dimensional charges are. We can never create an experiment to experience exactly what they are because we are trapped within our three dimensional perspective, just as the imaginary two dimensional creature was trapped within his two dimensional perspective upon the two dimensional plane.

The electron scattering experiments and the neutrino scattering experiments that were conducted in the 60's and 70's to determine the charge of the quarks only reacted with the "dotted line" on the quarks and not the quarks themselves. We only see their charges from our limited 3d perspective and not from the ultimate and higher dimensional perspectives of the universe. Consequently, the 1/3 & 2/3 chargers of quarks are not the actual charges on quarks.

Chapter 7
Creation of the ±1 Charge, the ±2 Charge and Spin

This chapter was originally part of the last chapter. However, because this topic is so fascinating it was decided to split the chapter into parts: For example: The theory called "Quantum Chromodynamics" is touted to be one of the most successful theories in all of science. Scientists praise it because its assignment of colors to quarks is used to explain the resultant charges on newly created particles after other particles collide and other particles decay. And yet, it has one major failing: try as it can, it cannot explain the most simple and Fundamental –1 charge of the electron, or the +1 Fundamental charge of the positron! And again, incredible as it may seem, the "supposedly" most successful theory used to explain the charges of quarks - cannot explain the simple The Fundamental Charge of nature!

QUARKS DO NOT CREATE THE CHARGE ON THE PARTICLE THEY INHABIT!!!

Although science presently believes that quarks create the ±1 charge on the particles they inhabit, this is a mistake! The –1 charge on any particle big or small is created by the same vortex of flowing space that creates the charge on the electron, and the +1 charge is created by the same vortex of flowing space that creates the charge on the positron. *Bigger "particles" are nothing more than enlarged electrons and positrons containing higher dimensional holes within them!* It happens like this:

When a disturbance upon the mutual 3d surface of the two volumes of space [such as two gamma rays colliding] pushes outward and into 4d space, the 4d space on SIDE 2, the contracting volume, is "pushed inward" into its interior because it is more elastic and more easily distorted inward than the expanding 4d space of SIDE 1. This push creates an indentation in SIDE 2 that 3d space begins to flow into and out of; becoming the two 3d holes called the electron and the positron.

Figure 7.1 Explaining the fundamental charge of nature:

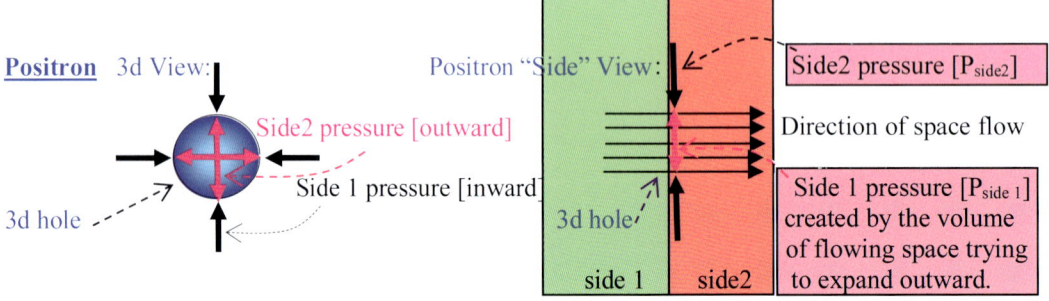

Consequently, the pressure (P) of these two sides reach a point of equilibrium where: $P_{side\ 1} = P_{side\ 2}$

As space flows into side 2 from side 1, the fundamental charge of nature [±1] is explained by the outward pressure of Side 2 equaling the inward pressure of the flowing space from side 1 that is trying to expand: allowing only a certain volume of 3d space to flow through the hole. These charges have to be the same throughout the universe. If not, the hydrogen atom would have many different spectral lines in different Galaxies [of different masses] and could not be identified.

Because we see the same spectral lines for hydrogen in all Galaxies, the volume of the flowing space has to be the same in all of their hydrogen atoms. Revealing a constant relationship between the pressure of side 1 and side 2; further confirming that the space throughout the universe is acting and reacting as a single entity: a single giant particle!

Figure 7.2

This figure represents a vortex formed when a region of the expanding volume of space [SIDE 1] pushed outward into the contracting volume of space [SIDE 2]. The ends of the vortex create the electron positron pair in this two dimension representation:

Note: This black line represents *three dimensional [3d] space:*

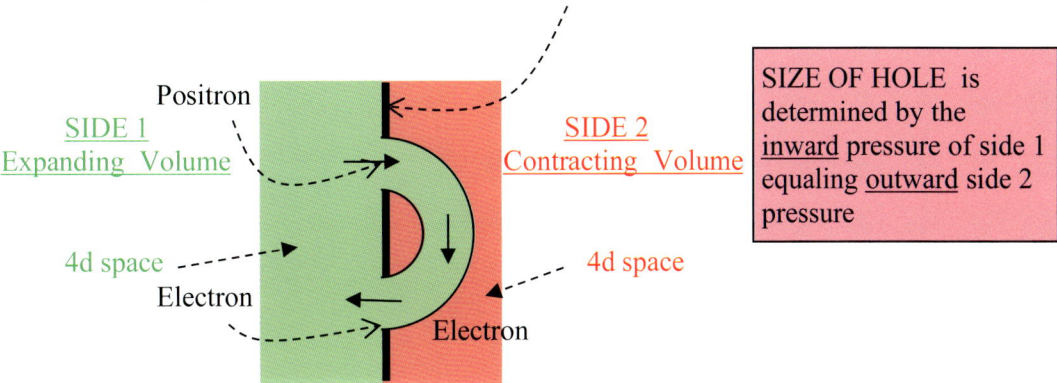

As seen in Figure 7.2, the charges on the electron and positron are created by the 3d space of SIDE 2 flowing into the positron, through a 4d vortex, then back into the electron, and finally outwards again onto the 3d surface of SIDE 2. However, if the initial collision is more energetic, the indentation continues on, into 5d space creating 4d holes upon its surface that are connected by 5d vortices. These 4d holes are the "particles" we call quarks; and what is important is that they exist within the volume of 4d space trapped within the confines of the enlarged 3d holes that were originally tiny electrons and positrons.

Because it is impossible to draw 4d holes, this concept can be illustrated using this two dimensional to three dimensional analogy:

Figure 7.3

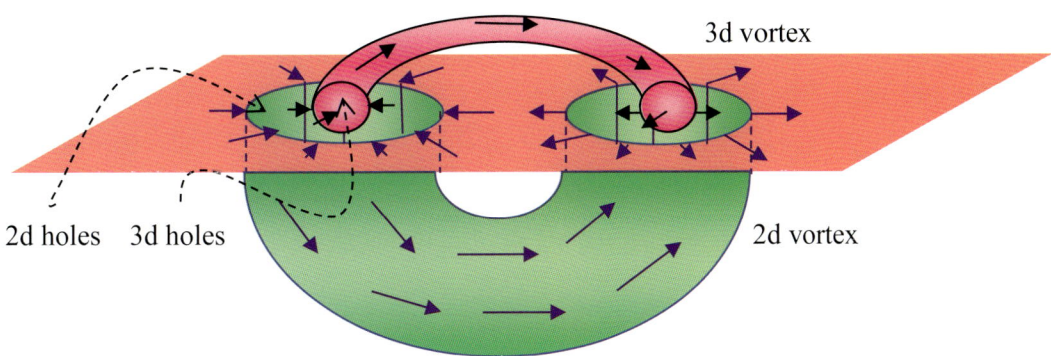

In Figure 7.3 notice how the three dimensional holes exist within the confines of the two dimensional holes. Notice how the two dimensional vortex of space flowing into and out of the two dimensional holes is really the two dimensional *surface* of the *three dimensional space it encloses*. And finally, notice how the three dimensional vortex that connects the three dimensional holes is "outside" of the two dimensional plane. And although it cannot be drawn, the three dimensional vortex is really the *surface* of the *fourth dimensional space it encloses.*

This same relationship exists between the vortices, the space particles are made of, and the quarks within them with one exception: the dimensions are all increased by one increment: the 2nd becomes the 3rd; the 3rd becomes the 4th; and the 4th becomes the 5th.

THE ±2 CHARGE OF A RESONANCE

Resonances are particles that possess ±2 charges. But how can this be possible? If all of the +1 charge's on particles are created by electron or positron vortices, how can a particle exist with a charge of ±2?

The answer is simple. Resonances are not one particle; instead, they are two particles attempting to combine to form one particle but failing. It happens like this:

Remembering that a particle is nothing more than the end of a vortex of flowing space, when two similar ends [ends that have the same charge] collide, the quarks within them try to recombine to form one single "particle." However, because the two ends of the vortices they exist within cannot combine to form one single end, the flowing space [their similar charges] within the two vortices repels them back apart. As they repel, the quarks recombine changing the quark content back to what it was before the collision. Hence, the same two original particles emerge from the collision.

The effect is analogous to two cars on the freeway colliding, locking bumpers for a short distance, and then separating. The key to understanding a resonance is the incredibly short life span of this unique "particle." The life span of a resonance is equivalent to the time it takes for the speed of light to traverse the distance across the nucleus of an atom!

Figure 7.4

| Example, two holes approach each other at high velocities: | the two holes collide and try to combine However, because the charges are similar, they cannot form one vortex. | Hence, their similar charges now force them back apart. |

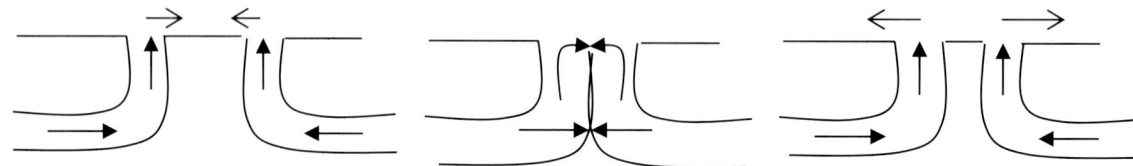

THE SPIN OF PARTICLES with quarks IS CREATED BY THEIR QUARK CONTENT

Although the charge of a particle containing quarks is created by its vortex of flowing space, its spin *is* created by its quark content. To understand how quarks create the spin of the particles they inhabit, let us examine two particles: the electron and the proton.

In *The Vortex Theory*, the explanation of intrinsic spin was presented. This explanation showed how a particle bent into a higher dimension of space can only spin in one of two orientations: clockwise or counterclockwise: called spin up or spin down. And this is precisely what is happening to the electron.

The electron is a 3d hole bent into 4d space. Within the 3d hole is a volume of 4d space that can only spin clockwise or counterclockwise. Hence, it possesses either spin up or spin down. The proton is entirely different.

As will be shown later, the proton is really a positron ballooned outward in size containing three quarks within the now greatly enlarged 4d volume of space within its interior. Because quarks are 4d holes upon the surface of 5d space, the 4d space within the interior of the proton is flowing into and out of these holes. As these three holes spin, the 4d volume of space within the interior spins. Since the quarks can only spin up or down, the spin of two quarks cancel each other and the proton takes on the spin of the third one.

For Pions and Kayons, which are "Particles" that only have two quarks, a different situation occurs. Since the 4d volume only contains two 4d holes, their spins either add or subtract making the total spin either one or zero.

Chapter 8
The True Vision of Space

> In Book 2, the Vortex Theory, the construction of space was explained. Because the construction of the universe requires the construction of space to be explained too, this chapter is a reproduction of the chapter of Book 2 explaining space. The explanation of the construction of space is the last piece of the puzzle necessary to explain the construction of the universe.
>
> Contrary to popular belief, space is not a void. Space is a multi-dimensional substance that everything in the three dimensional universe is made of. This idea is not a return to the old Aether theory. As we will soon see, the space within which we exist, is unlike anything anyone has ever imagined before.

The key to discovering how *everything* in the universe is constructed is found in the construction of space.

Contrary to present beliefs, space is not a void. Space is made of something. The substance that space is made of is totally unique from our point of view. It can both stretch and flow, is constructed out of at least seven dimensions, and is in a state of expansion.

Furthermore, our three dimensional universe is not infinite. The three dimensional universe only appears to be infinite from our perspective. The three dimensional space in which we exist is in reality the *finite* surface of a fourth dimensional volume of space.

Just as a two dimensional plane is the surface of a three dimensional object, the three dimensional space of our universe is the surface of fourth dimensional space; and fourth dimensional space is the surface of fifth dimensional space, etc. this is not supposition, it is based upon the work of two of the greatest mathematicians who ever lived: *Joseph-Louis Lagrange*, and *René Descartes*.

Hundreds of years ago, these two great men, amusing themselves by writing numbers on pieces of paper, and discovering new formulas, never realized the implications and consequences of what these formulas actually represented! For example:

It is easy to say, and no one will disagree, that we live in a vast, three dimensional volume of space. Everyone also will agree that because we live in three dimensions, the three components of three dimensional space called length, width and height, [identified mathematically as the three co-ordinates X, Y, & Z of the Cartesian Coordinate System] are subsequently used to define and identify the position of an object. However, is it just a co-incidence that a three dimensional volume of space just happens to be the surface area of a fourth dimensional sphere!

Although most people are aware of the fact that a sphere such as a basketball or a bowling ball of radius "R" has a two dimensional surface area: expressed mathematically as $4\pi R^2$; where the symbol R^2 = a two dimensional flat area. This two dimensional area encloses a three dimensional internal volume of space or matter, expressed mathematically as: $4/3\pi R^3$; where R^3 = the three dimensional volume of space.

Figure 8.1

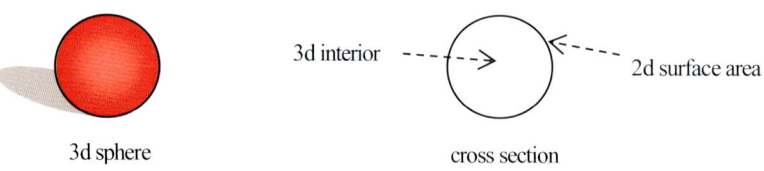

However, one of the great, *unrealized shocking truths of the universe* - discovered unwittingly by these two brilliant mathematicians hundreds of years ago - is the fact that a *three dimensional volume of* space is actually the *surface* area of a fourth dimensional object! Note: the *surface area* of a fourth dimensional sphere of radius "R" is equal to $2\pi^2 R^3$: where R^3 represents the fourth dimensional sphere's surface *area*! This *AREA* is in reality a *VOLUME* of three dimensional space! A three dimensional volume of space surrounding an internal fourth dimensional volume equal to $1/2\pi^2 R^4$. And again, it is important to realize that the *surface area* of the fourth dimensional sphere is *actually a three dimensional volume of space!!!*

Figure 8.2

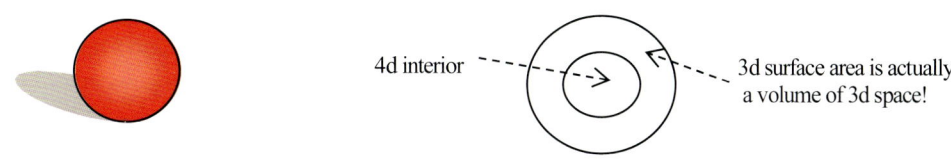

Schematic representation of a 4d sphere   Schematic representation of the cross section of a 4d sphere

What these mathematicians failed to realize is that these formulas reveal that our universe's three dimension volume of space is actually the surface area of a fourth dimensional object! This conclusion seems so outrageous that we immediately want to know if there is any other corroborating evidence to confirm this shocking conclusion? Is there?

The answer is yes, it starts like this…

As we look out into the universe with our most powerful telescopes, from our point of view, it appears we live in a vast three dimensional volume of space. However, there are problems with this vision. If everything that we see in the universe is a result of the Big Bang, where is its point of origin?

In 1054 AD, the people of the earth witnessed a supernova. Although they did not know what they were seeing, the Chinese wrote about a star that suddenly appeared and was so bright that it could be seen in the middle of the day; in the southwest, Native Americans drew stars on rocks in awe of this one of a kind event. Today we can witness the remnants of that explosion. By looking at the Crab Nebula and measuring the Doppler shift of its light, we can tell by the velocity of the particles streaming out into the universe that the nebula is the result of a gigantic explosion that occurred approximately 1000 years ago [approximately equaling 1054 AD]. We can also use the same technique to trace the coordinates of the particles streaming out into the universe from the Crab Nebula to determine the point of origin of the supernova. Hence, we can see approximately where the blast occurred.

So, how come we cannot do this with the Galaxies of the universe? How come we cannot use this same technique to determine the point of origin of the Galaxies created after the Big Bang?

Unfortunately, this same scientific method fails when we try to apply it to the motion of the Galaxies out of which the physical universe is constructed. Even more amazing, we find it hard to believe that when we look for the origin of the Big Bang, we discover that every point in the universe appears to be the point of origin we are looking for! How can this be?

Perhaps it can be best explained using this analogy: just like an imaginary microscopic creature living upon the 2d surface of an expanding balloon, because the entire surface is expanding at once, there are no beginning co-ordinates for anything upon its surface. Also, from his point of view, as he looks outward upon the surface of the sphere and rotates through 360 degrees, from his point of view, he appears to be standing upon the center of the sphere [the horizon appears to be equal distant in all directions he looks in]..

The same is true for us living upon this 3d surface and searching for the beginning of the Big Bang. What we label as the Big Bang was actually the point of origin for the beginning of the expansion of the 3d surface we live upon. Also, from our point of view, just like the imaginary creature living upon the 2d surface of the 3d sphere; no matter where we are appears from our perspective to be the center of our universe. Which forces us to solve another problem: where is the center of mass of the universe?

Every second semester engineering student knows the technique used to find the "Center of Mass" of an object. It is easily accomplished using a little Physics and Calculus. It is also used to determine the center of mass of planetary systems like ours, and the center of mass of Galaxies, and Galaxy clusters. So how come this same technique doesn't work when it comes to finding the center of mass of *all* the Galaxies existing in the universe?

And again, just like the imaginary though intelligent microscopic creatures living upon the surface of the balloon, find that they cannot measure the mass of the balloon; because unbeknownst to them, even though the material of the balloon exists upon its surface, the center of mass of the balloon resides in the empty space within the center of the balloon: and so it is with us. Even though from our perspective, we appear to live in a three dimensional volume of space, this space is really the surface of a higher dimensional volume of space. Consequently, the center of mass of all the Galaxies that exist in the universe lies within the *center of the fourth dimensional volume* and not upon the surface! Note: this same explanation for the center of mass still works if indeed fourth dimensional matter exists within a fifth dimensional volume of space [the center of mass will be in 5d space.]

Is there any more corroborating evidence? Again, the answer is yes; and brings us to the subject of infinity…!

Infinity is a problem most of us have tried yet failed to rationalize and reconcile within our minds. It is hard to imagine a never ending volume of space that goes on forever and ever. But that is what we appear to see. When we look out into the universe, it seems to go on forever and ever. Luckily, this is just another illusion.

Returning again to the perspective of intelligent microscopic creatures living upon the surface of the balloon, we find that they encounter the same mental problem: where is the end of the two dimensional surface they live upon? If the balloon is large enough, they find that if they go in search of its end, they can keep on traveling around and around it forever and ever, making them believe it is infinitely large. And this is the exact dilemma faced by us.

It appears from our point of view that the universe is never ending, that it goes on forever. But again, this problem is resolved when we realize this is merely an illusion, it is not real. If we could travel fast enough and far enough, we would come back to the exact same point we started out from! [Like Columbus, we could go west by going east!] Consequently, infinity is just another of the many illusions that are created from our limited point of view here within this three dimensional volume of the "surface area" of this fourth dimensional object!

Because many people reading this book are not scientists, a brief description of the dynamics of these higher dimensions of space is necessary. Unfortunately, because the fourth and fifth dimensions (called higher dimensional space for simplicity) are impossible to draw or describe, only analogies or inferences can be used to explain their unusual properties and qualities. It goes like this…

Roughly speaking, higher dimensional space is a volume of space located outside of our three dimensional universe. This volume of space has a unique perspective: an observer in three dimensional space cannot see anything that exists in higher dimensional space, however, from higher dimensional space, an observer can see everything that exists in three dimensional space!

The entrance into fourth dimensional space is also unique. To enter into fourth dimensional space, one must pass through three dimensional holes. Just as we must pass through two dimensional holes in three dimensional volume of space, we must pass through three dimensional holes to enter into the fourth dimensional volume of space (and fourth dimensional holes to enter fifth-dimensional space, etc.). Revealing that A Hole is Always One Dimension Less than the Number of Dimensions Present…

The relationship between the numbers of dimensions a hole is made of can be used to determine how many dimensions of space exist in the universe. Because a surface is always one dimension smaller than the volume it encloses, the entrance or hole leading into the interior of a volume is always one dimension smaller than the number of dimensions a volume is constructed out of.

For example, a characteristic indicating the existence of three dimensional space is found in the presence of two dimensional holes. It is easy to see that all holes in three dimensional space are two dimensional. All doors, windows, cave openings, pipe openings, and tunnel openings etc. are two dimensional holes.

The same is true for the existence of fourth dimensional space. If fourth dimensional space exists, a characteristic of its presence will be the existence of three dimensional holes within our physical universe. [We will find these holes in the next chapter.]

Unfortunately, great difficulty is encountered when first trying to visualize three dimensional holes and higher dimensions of space. This problem occurs because we both think and reason in terms of three dimensional space. Whereas, to think and reason in terms of fourth dimensional space, or fifth-dimensional space, we have to visualize the universe in a way that is totally different from anything we have ever done before.

A way to envision the relationship between the different dimensions of space is found in the following drawings:

Figure 8.3

One-dimensional space consists of one line.

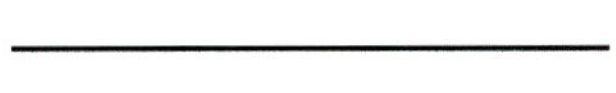

Figure 8.4

Two dimensional space is a plane that is at right angles to the line.

Figure 8.5

Three dimensional space is at right angles to the plane.

Figure 8.6

Fourth dimensional space is at right angles to three dimensional space. Since fourth dimensional space is impossible to draw, the following is only a representation, a schematic drawing.

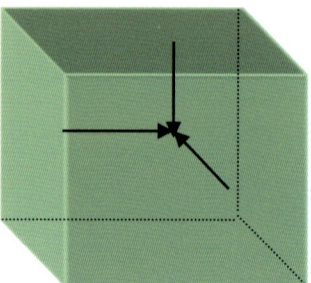

Another way to understand this construction is to look up at a corner of the room you are sitting in. Observe the point where the ceiling and the two walls meet. The ceiling and the two walls represent the three dimensions of space we live in: length, width, and height. Now observe how the ceiling is at right angles to both walls, while both walls are at right angles to the ceiling and the opposite wall. This relationship allows us to observe the fact that each dimension is at right angles to the other two dimensions *simultaneously*.

Although the ninety-degree angles that exist between the three dimensions and create three dimensional space are easy to see, the ninety-degree angle that exists between fourth dimensional space and the three lower dimensions is impossible to visualize.

Even though we cannot visualize the ninety-degree angle that exists between three dimensional space and fourth dimensional space, what we can do is to visualize what the entrance into fourth dimensional space will look like from our three dimensional perspective. Because this entrance is a three dimensional hole, it will be spherical in shape; and if it is small - very small - it will appear to be a tiny spherical particle!

Figure 8.7

23

Chapter 9
Creation and Destruction of the Universe

At last! We now possess the knowledge necessary to be able to explain the creation of the universe as it is today. Although it is impossible to draw higher dimensional space, these 2d to 3d schematic drawings are used.

The knowledge presented in previous chapters are contained within the drawings below:

Quarks allow us to realize that higher dimensional space has to exist. The discovery of 1/3 and 2/3 charges allow us to realize that two volumes of higher dimensional space exist; with two different elasticity's created by one volume in expansion, one in contraction. The two sides of space also reveal that our three dimensional space is trapped between both volumes. And the mathematics of three dimensional space reveals that it is the surface of fourth dimensional space.

THE SCHEMATIC REPRESENTATION OF THE ONE UNIVERSAL "PARTICLE"

The internal green circular ring below represents dimensions 4 - 7 on the *expanding side*; our lavender third expanding dimension is in the middle; and red contracting dimensions 4 - 7 are on the *contracting side* of space. The entire drawing represents one massive particle.

Figure 9.1

green = volume in expansion
red = volume in contraction
violet = our 3d volume of space
[Not to proportion]

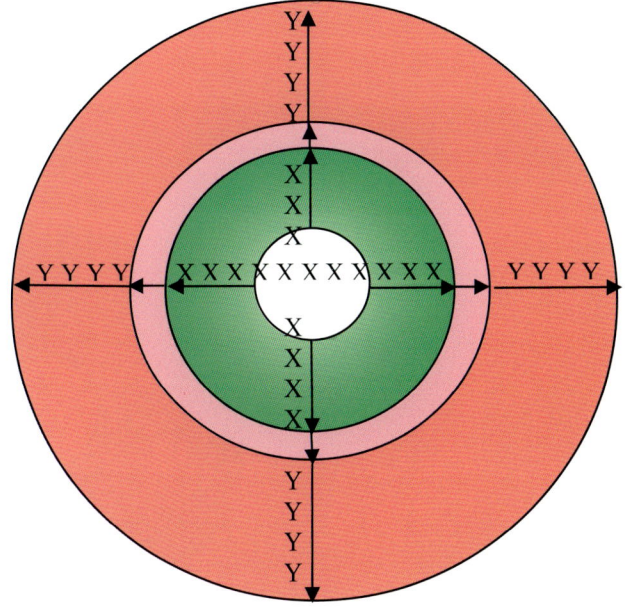

Note: the 3d volume [violet] is also expanding because one side [red] is contracting while the other side [green] is expanding.

24

Figure 9.2 Side view

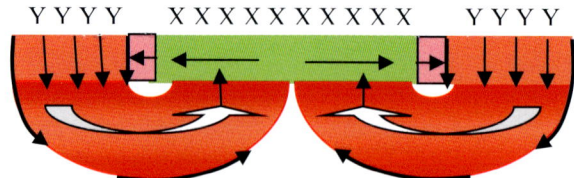

In the above two drawings, the "Y" letters represents the region where higher dimensional space is *contracting* in volume as it turns inside out and *adds* to the expanding green volume. The twisted white arrows reveal that this red volume of space is turning inside out.

The "X" letters represent the expanding region of higher dimensional space that is growing larger, increasing in volume as the contracting volume flows into it.

It also must be understood that Figure 9.2 is a cross section side view of a "three dimensional donut" of space flowing into the green volume [expanding volume].

WHAT IS THE FATE – THE DESTINY OF OUR UNIVERSE?

The expanding area will continue to expand, and the contracting area will continue to contract as our three dimensional surface [violet circle] also continues to increase in size between these two volumes.

Figure 9.3

green = volume in expansion
red = volume in contraction
violet = our 3d volume of space
[Not to proportion.]

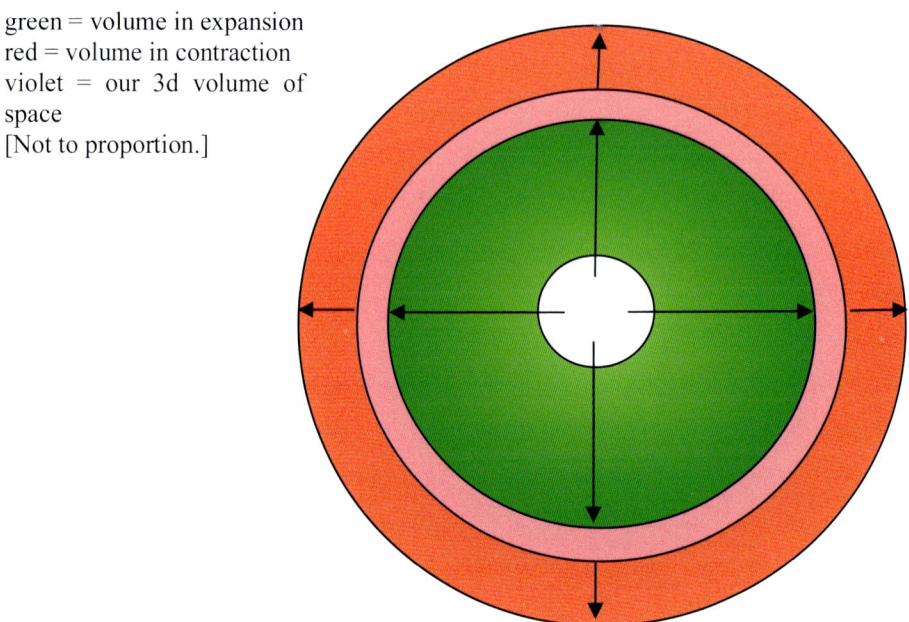

When the expansion is completed, the three dimensional surface will be on the very edge of fourth dimensional space [Figure 9.4]. At this instant, the contracting volume will be gone and all of the three dimensional vortices that formed on the contracting fourth dimensional volume will disappear in the blink of an eye! All matter will instantly disappear. All that will be left are the regions of denser and less dense space on the expanding higher dimensional side of space. [Figure 9.5]. However, this is not the end, but a new beginning!

Note the arrows that indicate the expanding volume! Because the contracting volume has disappeared, and no more space is being added to the expanding volume [that is still in expansion]; consequently, at the geometric center of the expanding volume, a rip in space will occur [Figure 9.6]; at this instant, the outward edge of the expanding volume of space will curve back and begin to fill the void in the center. Two new volumes will reoccur: an expanding volume and a contracting volume; and the whole process will begin all over again as seen in [Figure 9.7].

Figure 9.4   **Figure 9.5**

Figure 9.6   **Figure 9.7**

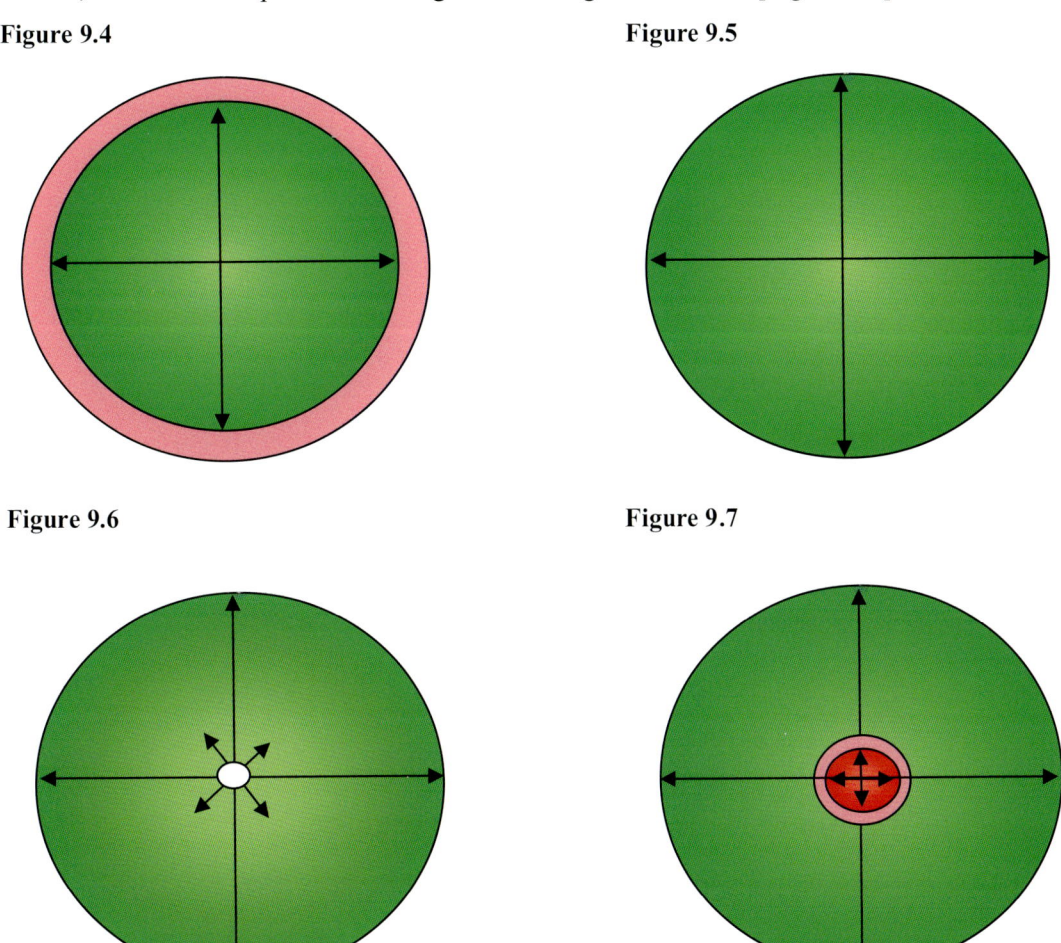

The above scenario of the universe dying and then being reborn again is not speculation. It is based upon the strange and wide variety of galaxy combinations. For example, if the creation of matter upon the expanding volume was uniform, the placement of galaxies in the universe would also be uniform. But they are not! Instead there are a huge variety of combinations: everywhere we look, there are walls of galaxies, globs of galaxies, long strings of galaxies, and many small clusters of galaxies.

This creation of these bizarre diverse conglomerations of galaxies can be explained like this: at the start of a new cycle, if when the expanding volume begins to add to the contracting volume, and if there are denser regions of space that had no galaxies in it, and if there are less dense regions where galaxies were; as these less dense regions begin to flow into the new expanding volume, matter will find it easier to reform matter there; and avoid the denser regions of space where the vortices of matter are harder to form. Causing and creating galaxy formations that are no longer uniformly spaced, but instead, reconforming to previous arrangements.

CURRENT SCIENTIFIC BELIEFS IN THE UNIVERSE'S CREATION AND CONSTRUCTION

According to 20th Century Science, the universe began with what is called the Big Bang; and everything, everything that exists in the universe today was concentrated into one tiny, microscopic dot called the Singularity! This Singularity subsequently exploded creating the Big Bang.

Figure 9.8

The Singularity

Figure 9.9 The Singularity explodes creating the Big Bang

3d space [Space-time]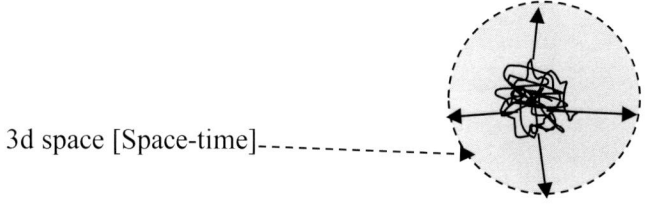

Problems with this idea:

1. Where did the Singularity come from? [This is a question nobody has ever answered!]

2. What force caused it to blow up? Was this some new "one time" force?

Note, 20th Century Science's explanation of the big bang lists the creation of the forces of nature as coming *after* the big bang occurred. So, what force caused the singularity to blow up?

3. Where did the fourth dimension of "Spacetime" come from?

Cosmologists now try to state that this fourth dimension of Spacetime was "wrapped up" in the singularity! But this is an "Ad Hock," contrived explanation made for the convenience of the theorists who have no explanation for it. They don't know where it came from; so to eliminate arguments, these confused scientists added it to the singularity: and this creates another big problem! Because, according to 20th Century science, Space-time is a volume of space made out of a REAL substance called "Time" [there are five pieces of the universe: matter, space, time, energy and the forces of nature]! According to Albert Einstein, spacetime is responsible for creating the "Length Shrinkage" and "Time Dilation" effects that take place at near light velocities. But if so, then "Time" has to be something REAL: a real "substance" creating real effects on matter that is also made of something! [If spacetime is made of nothing, then, how can "nothing" cause matter to shrink?

In conclusion, this convoluted vision of the universe that has three dimensions made of nothing and one made of something is ludicrous!

4. Where is the Point of Origin of the big Bang?

Another great mistake in this absurd hypothesis of the Singularity, and its subsequent massive explosion called the Big Bang was a question asked in the last chapter; where is the point of origin of the Big Bang? Unfortunately, there is no evidence in astronomy to indicate a point of origin! In fact, and as previously explained in Book 2, every point in the universe appears to be the center of the universe: revealing that the three dimensional space we live in is not a volume, but instead, is the surface of a fourth dimensional object!

Chapter 10
Creation of the Up and Down Quarks

> The Vortex Theory's explanation of the creation of the universe is reaffirmed again and again in the explanation of the particles of the universe. It is exciting to see how the creation of the up and down quarks is a validation that there are two volumes of <u>fourth dimensional space</u>. Furthermore, although it is impossible to visualize, not only do each of these two volumes touch the third dimension, they also touch each other. If not, these two different volumes of space could not decay into each other as will later be shown.

THE UP AND ANTI-UP QUARKS

Remembering that each lower dimension of space is actually the surface of the next higher dimension of space, it is now easy to understand that the up and anti-up quarks are fourth dimensional holes created within the fourth dimensional volume of space trapped within three dimensional holes. Furthermore, these up and anti-up quarks are created upon SIDE 1 of the fourth dimensional surface but who's vortex flows into SIDE 2 [as seen below in Figure 10.1.]

The three dimensional holes in space are enlarged electrons or positrons. Since up and anti-up quarks are fourth dimensional holes existing upon the surface of fifth dimensional space, up and anti-up quarks can be considered to be a *second layer of matter*: a hole within a hole.

The up quark possesses a +2/3 charge because of the difference in the elasticity between SIDE 2 and SIDE 1. [Again, if it is twice as easy for space to flow from SIDE 1 into SIDE 2 than it is to flow from SIDE 2 into SIDE 1, the volume of a SIDE 1 hole will be twice the volume of a SIDE 2 hole. Hence, if a SIDE 1 hole's charge is 2/3, a SIDE 2 hole's charge will be 1/3.]

When up quarks are created in high energy collisions, because an up quark is really just one end of a vortex of fourth dimensional flowing space, the anti-up quark is formed along with the up quark. The anti-up quark has a charge of -2/3 because space flows back onto the 4d surface and is at the other end of the same 5d vortex that creates the up quark. This relationship can be seen in below in Figure 10.1:

Figure 10.1

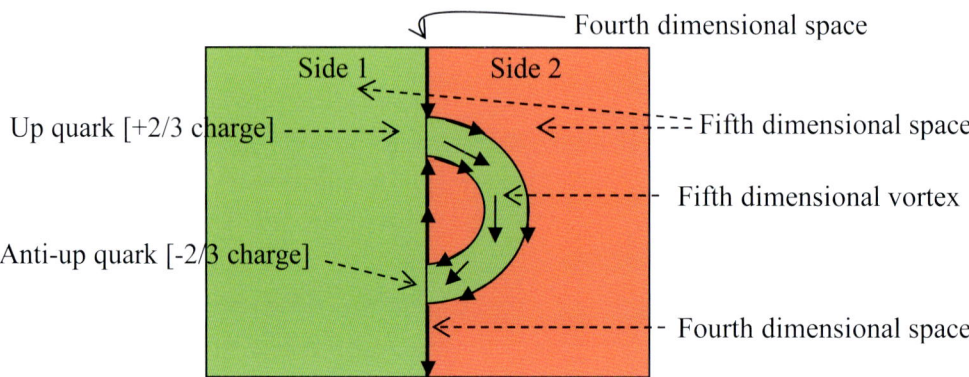

In the above drawing, the dark black line between the two volumes made of arrows represents the fourth dimensional surface of the fifth dimensional volume of space that the up and anti-up quarks are formed upon. The arrows on the dark line represent the fourth dimensional charge flowing from

29

the fourth dimensional surface into the up quark, through 5d space, and then exiting anti-up quark to flow back onto the fourth dimensional surface. To the left of the line is SIDE 1 – the expanding volume space; while to the right is SIDE 2 – the contracting volume of space. It should also be noted that the fourth dimensional line is also a line of demarcation separating the two volumes of *fifth dimensional space* to either side of the line.

The arrows <u>within</u> the vortex represent the direction the <u>volume</u> of *fifth dimensional space* is flowing. Note too, that in the up quark, 4d space is flowing off of the four dimensional plane and into the 5d volume; while in the anti-up quark, this same 4d space is returning back to the 4d plane.

SYMBOLS USED TO IDENTIFY UP AND ANTI-UP QUARKS

The following symbols are used to identify Up and Anti-up quarks:

Figure 10.2

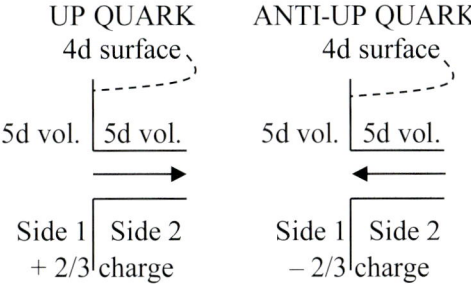

THE DOWN QUARK

The down quark is a fourth dimensional hole created upon SIDE 2 of the fourth dimensional surface of fifth dimensional space. Like the up quark, the down quark is also a fourth dimensional hole created within a preexisting three dimensional hole; so, just like the up quarks, the down quarks can be considered to be a second layer of matter [a hole within a hole].

The down quark possesses a -1/3 charge because its cross-section area is one half the cross-sectional area of the up quark.

Figure 10.3

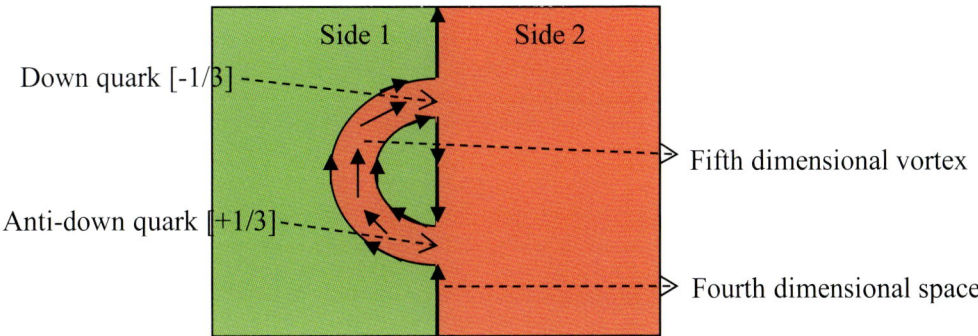

And just like the up anti-up quarks, when down quarks are created in high energy collisions, because a down quark is really just one end of a vortex of fifth dimensional flowing space, when initially created – via the collision of two particles – the anti-down quark is formed with the down quark: affirming the law of associated production.

In the above drawing, as with the up quark, the thick black line represents the fourth dimensional surface of the fifth dimensional volume of space that the down and anti-down quarks are formed upon. The arrows on the line represent the fourth dimensional charge flowing into the anti-down quark and out of the down quark. To the left of the line is SIDE 1 of fourth dimensional space, while to the right is SIDE 2 of fourth dimensional space.

The arrows in the vortex represent the direction fifth dimensional space is flowing. Note that in the anti-down quark, space is flowing off of the four dimensional plane, while in the down quark, the four dimensional space is returning to the four dimensional plane. This flow is the exact opposite to the flows in the up and anti-up quarks. However, their signs correspond to the direction space flows onto or off of the fourth dimensional plane.

SYMBOLS USED TO IDENTIFY DOWN AND ANTI-DOWN QUARKS

The following symbols are used to identify Down and Anti-Down quarks:

Figure 10.4

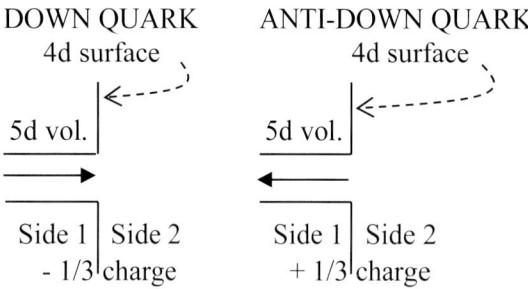

Chapter 11
The Creation of the Strange and Charm Quarks

> The creation of the strange and charm quarks are the first indication that the two volumes of space – one in expansion, one in contraction – each are made out of at least 5 dimensions of space. Furthermore, [and again], although it is impossible to visualize, not only do each of these 5d volumes touch the fourth dimension, they also touch each other. If not, they could not decay into each other as will later be shown.

THE STRANGE QUARK

The strange quark is a fifth dimensional hole created upon SIDE 2 of the *fifth dimensional surface of sixth dimensional space*. This fifth dimensional hole is created within a fourth dimensional hole – the down or anti-down quark. As such, a strange quark can be considered to be a third layer of matter: a hole within a hole, within a hole!

Figure 11.1 Strange quarks as ends of vortices:

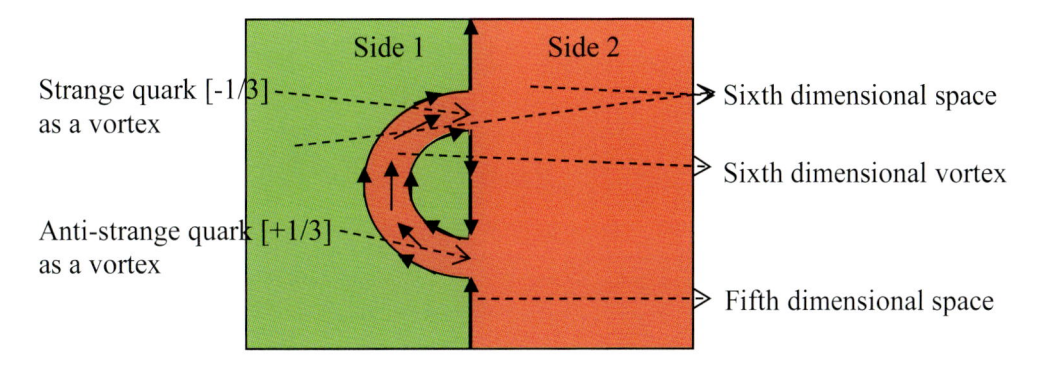

In the above drawing, the thick dark line represents the 5d surface of the 6d volume of space that the strange and anti-strange quarks are formed upon. The arrows on the line represent the 5d charge flowing into the anti-strange quark and out of the strange quark. To the left of the line is SIDE 1 of 6d space, while to the right is SIDE 2 of 6d space. The arrows in the vortex represent the direction 6d space is flowing. Note that in the anti-strange quark, space is flowing off of the 5d plane, while in the strange quark, the 5d space is returning to the 5d plane.

THE EXPLANATION OF THE CONSERVATION OF STRANGENESS

It was noted among the remnants of particle collisions that some of the newly created particles take a long time to decay. Two quark particles such as Kayons containing strange quarks seem to take a long time to decay in reference to two quark particles not containing quarks - such as Pions. This phenomenon appears to be a result of the fact that like the up and down quarks, the strange quark's mass is also light. Consequently, unlike the heavier quarks – the charm, top & bottom quarks whose mass causes the surrounding space to "squeeze" them quickly, causing them to decay quickly – the strange quark's lighter mass causes the surrounding space to "squeeze" them slowly; making them have longer lives.

The strange quark also possesses a -1/3 charge because it is sheathed inside of a down quark whose charge is -1/3: an observation that suddenly creates an enormous problem for measuring the true charge of the strange quark. Because unless particle accelerator technology vastly improves far beyond anything it is capable of doing today, there is no "particle" small enough to penetrate the interior of the strange quark to see exactly just what its charge really is!

Unlike the electron scattering experiments conducted at Stanford used to penetrate the interior of nucleons, we may never be able to penetrate the interior of the Down quark to experimentally test the charge on the Strange quark.

SYMBOLS USED TO IDENTIFY STRANGE AND ANTI-STRANGE QUARKS

Figure 11.2

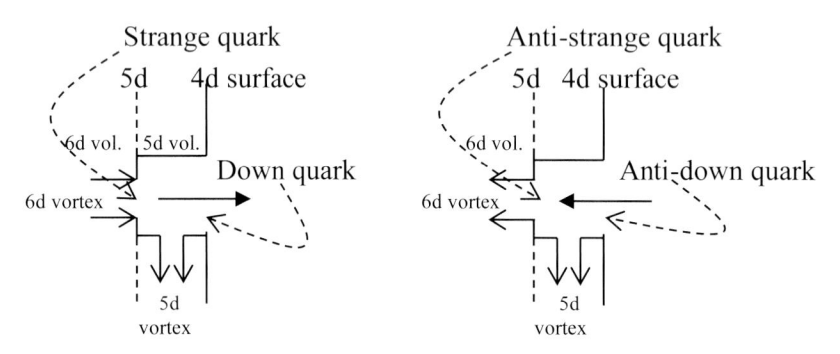

Note how both the strange and anti-strange quarks are formed upon the surface of 5d space *within* down and anti-down quarks, and that their vortices flow into and out of the volume of 6d space. Notice too how both the strange and down quarks have their own separate vortices, and how the anti-strange and anti-down quarks also have their own separate vortices. And again, because we cannot penetrate into the interior of the down quark, the strange quarks charge appears to be a value of 1/3.

THE CHARM QUARK

The charm quark is a fifth dimensional hole created upon SIDE 1 of the *fifth dimensional surface of sixth dimensional space*. This fifth dimensional hole is created within a preexisting fourth dimensional hole [the up quark]. Like the strange quark, the charm quark can also be considered to be a third layer of matter: a hole within a hole within a hole!

The charm quark possesses a +2/3 charge because it is sheathed within an up quark whose charge is +2/3. And like the strange quark, the true charge of the charm quark is impossible to measure.

Again, when quarks are created in high energy collisions, because a charm quark is really just one end of a vortex of sixth dimensional flowing space, when initially created via the collision of two particles within a linear accelerator, the anti-charm quark is also formed along with the charm quark. [The anti-charm quark resides within the anti-up quark.]

Figure 11.3

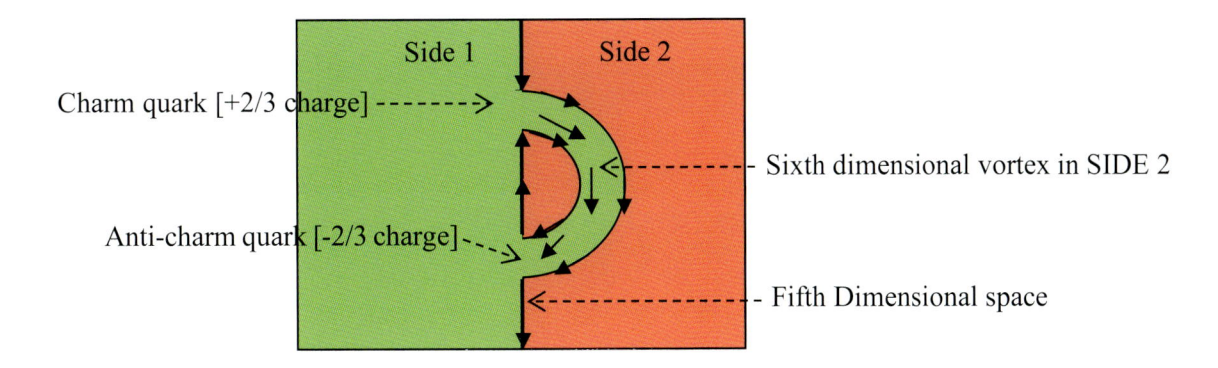

[Diagram: Side 1 (green) and Side 2 (orange) with labels:
- *Charm quark [+2/3 charge] →*
- *Sixth dimensional vortex in SIDE 2*
- *Anti-charm quark [-2/3 charge] →*
- *Fifth Dimensional space]*

In the above drawing, the thick dark black line made of arrows represents the 5d surface of the 6d volume of space that the charm and anti-charm quarks are formed upon. The arrows on the dark line represent the 5d charge flowing from the 5d surface into the charm quark, through 6d space, and then exiting the anti-charm quark to flow back onto the 5d surface. The arrows within the vortex represent the direction that the volume that the *sixth dimension of space* is flowing in.

To the left of the dark line is SIDE 1: the expanding volume space. While to the right is SIDE 2: the contracting volume of space. It should also be noted that the 5d line is also a line of demarcation separating the two volumes of *sixth dimensional* contracting and expanding space to either side of the line.

SYMBOLS USED TO IDENTIFY CHARM AND ANTI-CHARM QUARKS

Figure 11.4

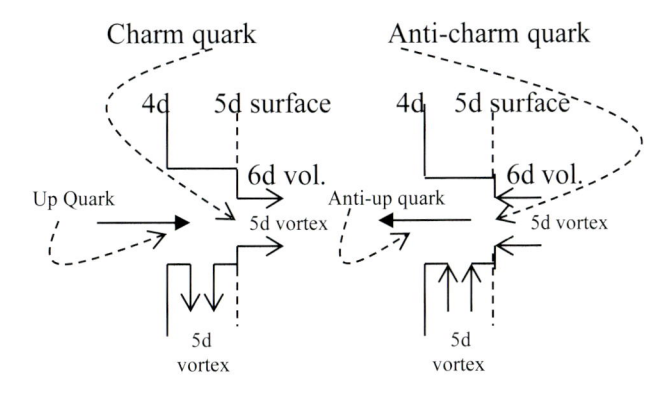

[Diagram showing Charm quark and Anti-charm quark symbols with labels: 4d, 5d surface, 6d vol., Up Quark, Anti-up quark, 5d vortex]

Note how both the charm and anti-charm quarks are formed upon the surface of 5d space within up and anti-up quarks, and that their vortices flow into and out of the 6d volume of space. Note too, that their positive or negative 2/3 charges are a result of the up and anti-up quarks they reside within. Their actual charges cannot be determined at this moment in history due to the fact that we do not possess the more advanced equipment necessary to do the tests.

Chapter 12
Creation of the Top and Bottom Quarks

The creation of the top and bottom quarks is the first indication that the two volumes of expanding and contracting space each possess at least <u>seven dimensions of space</u>. Furthermore, [and again], although it is impossible to visualize, not only do each of these 7d volumes touch the sixth dimension, they also touch each other. If not, they could not decay back and forth into each other as will be shown later.

THE BOTTOM QUARK

The bottom quark is a sixth dimensional hole created upon SIDE 2 of the *sixth dimensional surface of seventh dimensional space*. This sixth dimensional hole is created within a 5d hole: [the strange quark]. As such, a bottom quark can be considered to be a fourth layer of matter: a hole, within a hole, within a hole, within a hole!

The bottom quark possesses a -1/3 charge because it is located within a strange quark within a down quark whose charge is -1/3. The charge on the bottom quark cannot be determined by experimentation!

And again, when quarks are created in high energy collisions, because a bottom quark is really just one end of a vortex of seventh dimensional flowing space, when initially created – via the collision of two particles – the anti-bottom quark is formed with the bottom quark. [The anti-bottom quark is located within the anti-strange quark.]

Figure 12.1

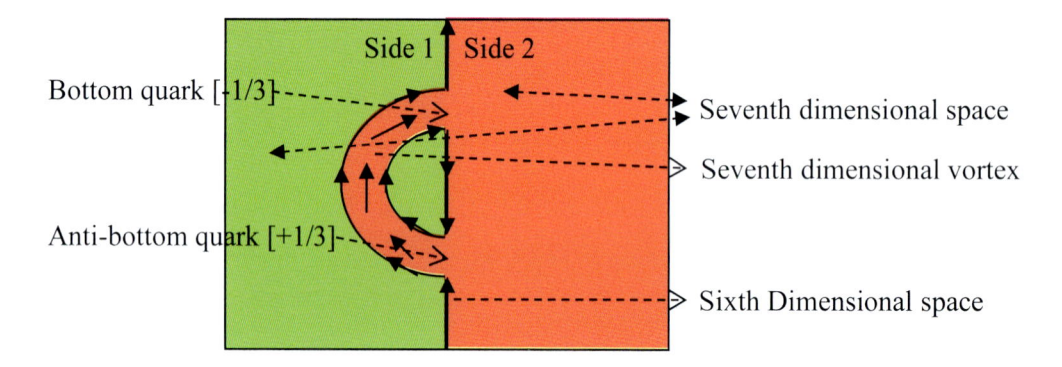

In the above drawing, the thick dark line represents the sixth dimensional surface of the seventh dimensional volume of space that the bottom and anti-bottom quarks are formed upon. The arrows on the line represent the sixth dimensional charge flowing into the anti-bottom quark and out of the bottom quark. To the left of the line is the six dimensional space of SIDE 1, while to the right is the six dimensional space of SIDE 2. The arrows in the vortex represent the direction seventh dimensional space is flowing. Note that in the anti-bottom quark, space is flowing off of the sixth dimensional plane, while in the bottom quark, the sixth dimensional space is returning to the sixth dimensional plane. Again, this flow is the exact opposite to the flows in the up and anti-up quarks.

SYMBOLS USED TO IDENTIFY BOTTOM AND ANTI-BOTTOM QUARKS

Figure 12.2

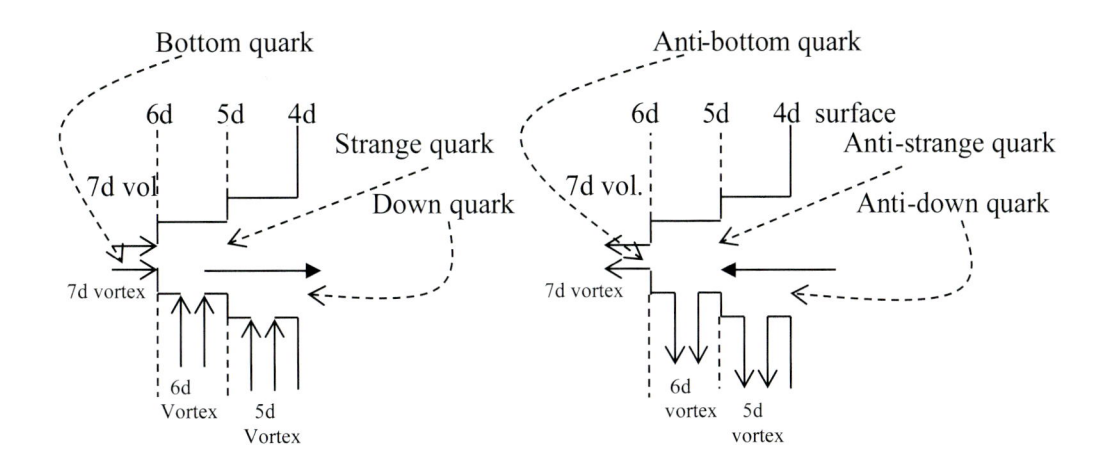

Note how the bottom and anti-bottom quarks are formed within strange and anti-strange quarks on side 1 of space and upon the 6d surface; and that their vortices flow into and out of 7d space. Notice too how the Bottom, Strange, Down, Anti-bottom, Anti-strange, and the Anti-down all possess their own vortices.

THE TOP QUARK

The top quark is a sixth dimensional hole created upon SIDE 1 of the *sixth dimensional surface of seventh dimensional space*. This sixth dimensional hole is created within a preexisting fifth dimensional hole [the charm or anti-charm quark]. Like the bottom quark, the top quark can also be considered to be a fourth layer of matter: a hole within a hole, within a hole, within a hole!

The top quark possesses a +2/3 charge because it is located within a charm quark. Like the bottom quark, it is technically impossible at this point in history to determine just what its charge is.

Again, when quarks are created in high energy collisions, because a top quark is really just one end of a vortex of seventh dimensional flowing space, when initially created via the collision of two particles, the anti-top quark is formed along with the top quark. [The anti-top quark resides within the anti-charm quark.]

Figure 12.3

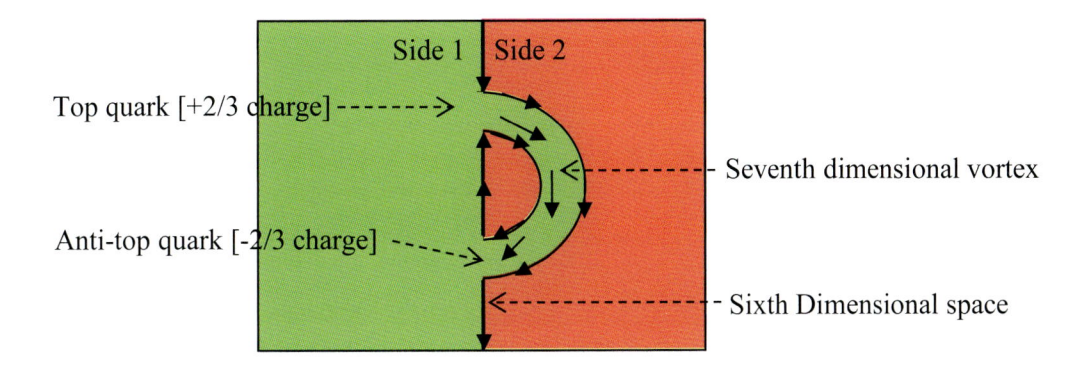

As with the bottom quark, in the above drawing, the thick dark line represents the 6d surface of the 7d volume of space that the top and anti-top quarks are formed upon. The arrows on the line represent the 6d charge flowing into the top quark and out of the anti-top quark. To the left of the line is SIDE 1 of 7d space, while to the right is SIDE 2 of 7d space. The arrows in the vortex represent the direction 7d space is flowing. Note that in the top quark, space is flowing off of the 6d plane, while in the anti-top quark, the 6d space is returning to the 6d plane.

SYMBOLS USED TO IDENTIFY TOP AND ANTI-TOP QUARKS

Figure 12.4

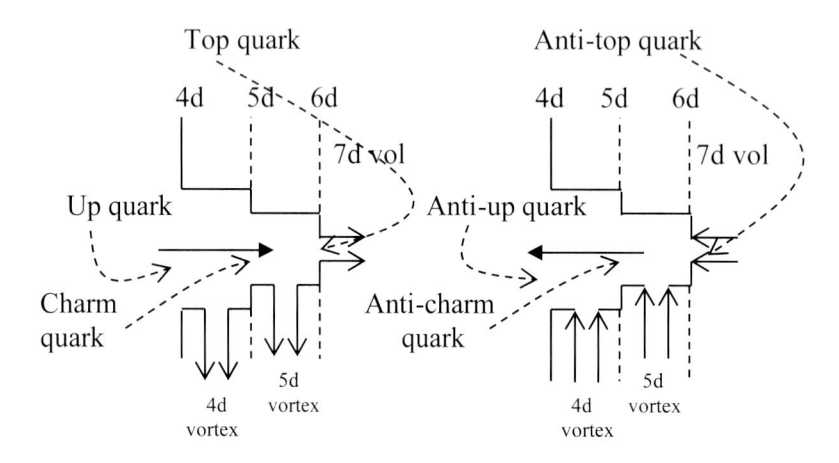

Note how the top and anti-top quarks are formed within charm and anti-charm quarks on side 2 of space and upon the 6d surface and that their vortices flow into and out of 7d space. Notice too how the Top, Charm, Up, Anti-top, Anti-charm, and Anti-up quarks all have their own vortices.

SCHEMATIC DRAWING SHOWING HOW EACH HIGHER DIMENSION TOUCHES EACH OPPOSITE DIMENSION AND ITS LOWER DIMENSION:

As before, the red shapes represent dimensions in contraction, while the green shapes represent dimensions in expansion. Although it is a bizarre drawing, it nevertheless allows us to picture in our mind a way to see how a dimension can touch a lower dimension, a higher dimension, and at the same instant, touch its opposite dimension in contraction [or expansion].

Figure 12.5

For example: note how the 6th dimension in expansion [green] touches both the 5th dimension in expansion and the 6th dimension in contraction [red] and the 7th dimension in expansion. Also, in the figure below, the top is seen as a hole within the charm [and the charm is a hole within the charm [and the charm is a hole within the up.] So, when the top decays, pushing side 1 into side 2 creating the bottom quark, on Side 1 leaving the charm on Side 2.

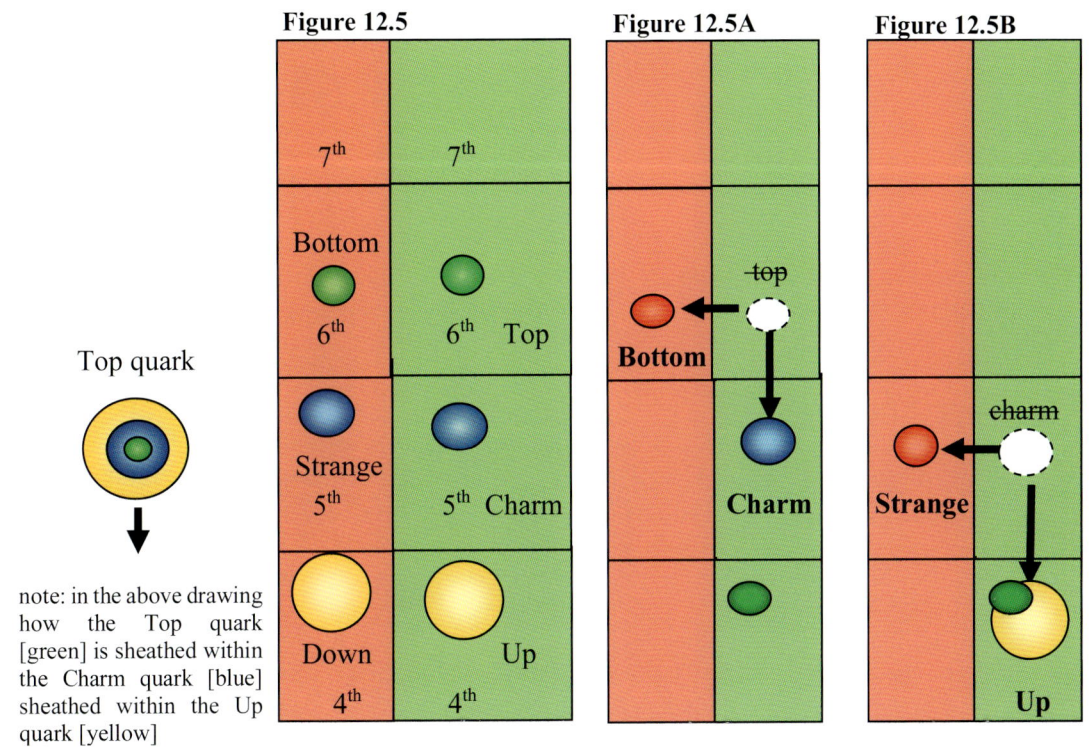

Top quark

note: in the above drawing how the Top quark [green] is sheathed within the Charm quark [blue] sheathed within the Up quark [yellow]

Consequently, it can now be seen that in some decay modes how the **Top** in Figure 12.5A will decay into the **Bottom** and the **Charm** [it was sheathed within]; then in Figure 12.5B how the **Charm** can decay into a **Strange** and the **Up** [it was sheathed within].

This schematic seen in Figure 12.5A above shows the decay mode of a top quark pushing out into side 1, creating a bottom quark while leaving the charm quark it was sheathed within.

38

Chapter 13
"The Four Layers of Matter"

> If it were not for the PhD thesis presented at the end of Book 1, *THE VORTEX THEORY OF ATOMIC PARTICLES*, we would never have suspected that "particles" such as protons and electrons are actually holes in space. Also, it would never have been suspected that the quarks existing within protons could also be holes: holes within holes; and, returning to the original hypothesis at the beginning of this book, we would not have suspected that quarks are really a hierarchy of holes and revealing the existence of two volumes of higher dimensional space:

Figure 13.1

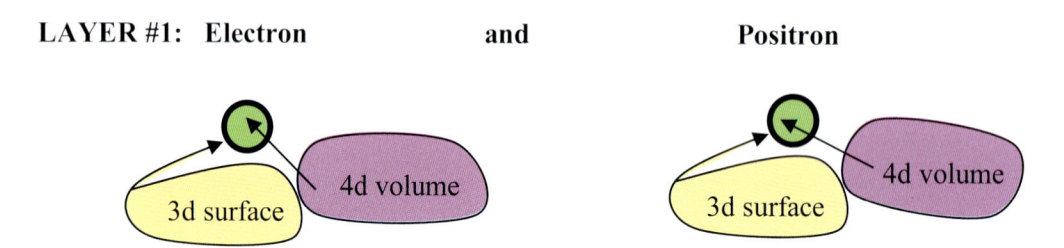

LAYER #1:   Electron   and   Positron

[Note how the electron and positron are 3d holes within 4d space. Both the electron and positron are colored green because they are both formed on side 1 of space: the expanding volume.

Figure 13.2

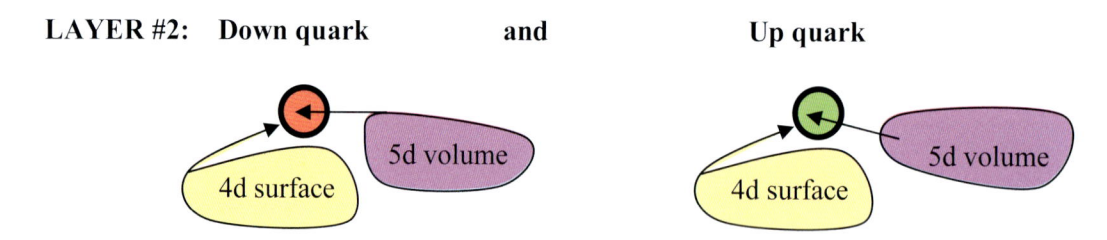

LAYER #2:   Down quark   and   Up quark

[Note how the up and down quarks are 4d holes within 5d space. The down quark is red because of its side 2 construction; the up quark is green because of its side 1 construction.

39

Figure 13.3

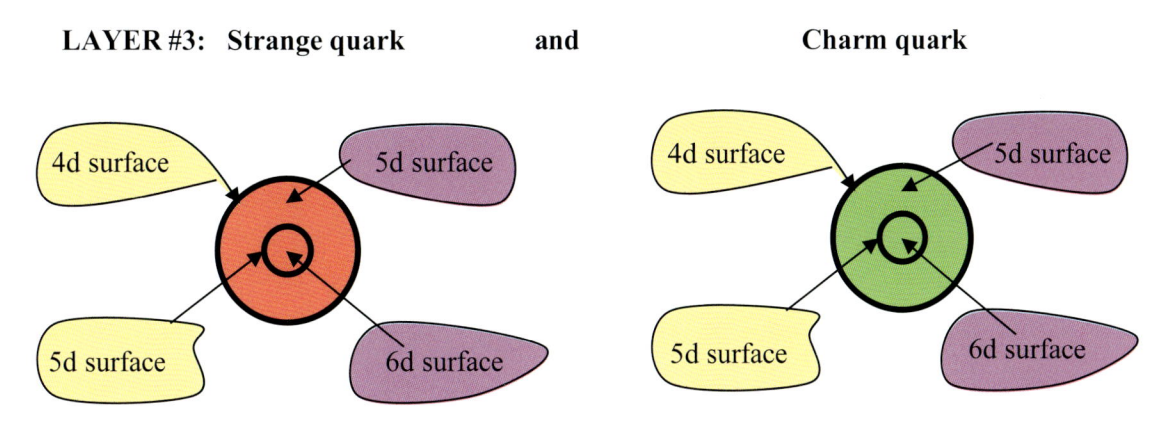

LAYER #3:   Strange quark   and   Charm quark

[Note how both the strange and charm quarks are 5d holes within 4d holes. Also, because they are located within down and up quarks they possess the 1/3 and 2/3 charges of these lower dimensional holes they are "sheathed" within. (Strange is side 2; while charm is side 1)]

Figure 13.4

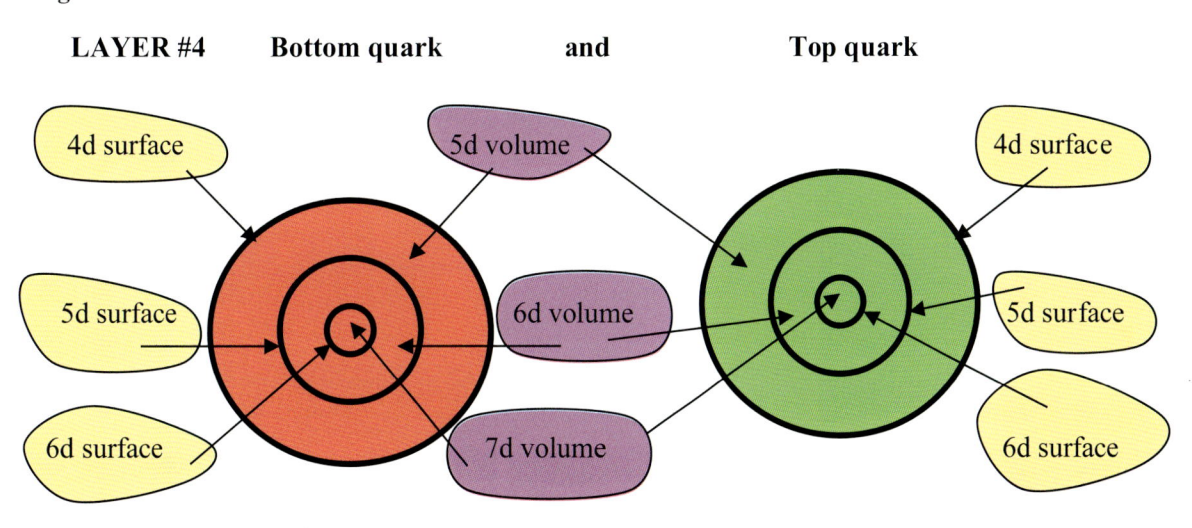

LAYER #4   Bottom quark   and   Top quark

Note how both the bottom and top quarks are 6d holes within 5d holes. Also, because they are located within down and up quarks they possess the 1/3 and 2/3 charges of these lower dimensional holes they are "sheathed" within. (Bottom is side 2; while top is side 1)]

Chapter 14
The Weak Force

> Although the weak force was explained in Book 2, the Vortex Theory, it must be explained here in greater depth before "TUNNELING" can be explained. Only when the Weak Force is properly understood is lepton decay properly understood. <u>The present 20th Century explanation of the Weak force using W & Z bosons is a mistake!</u>
>
> Unlike the other forces of nature, the true explanation of the Weak Force is another affirmation of the original hypothesis presented at the start of this book, that the universe is constructed out of one single massive multi-dimensional particle turning inside out creating two sides of space sharing a mutual surface: it does not need boson particle to mediate force!

Although it is currently believed that the W particle and the Z particle transmit the weak force, this is a mistake. The W particle and the Z particle are only manifestations of unfolding vortices and do "mediate" the weak force: *nor do they create the weak force.*

The weak force is created by the difference in the elasticity's of the two sides of space being out of balance. This balance is best demonstrated below using a quark anti-quark pair.

Figure 14.1
"SIDE VIEW"

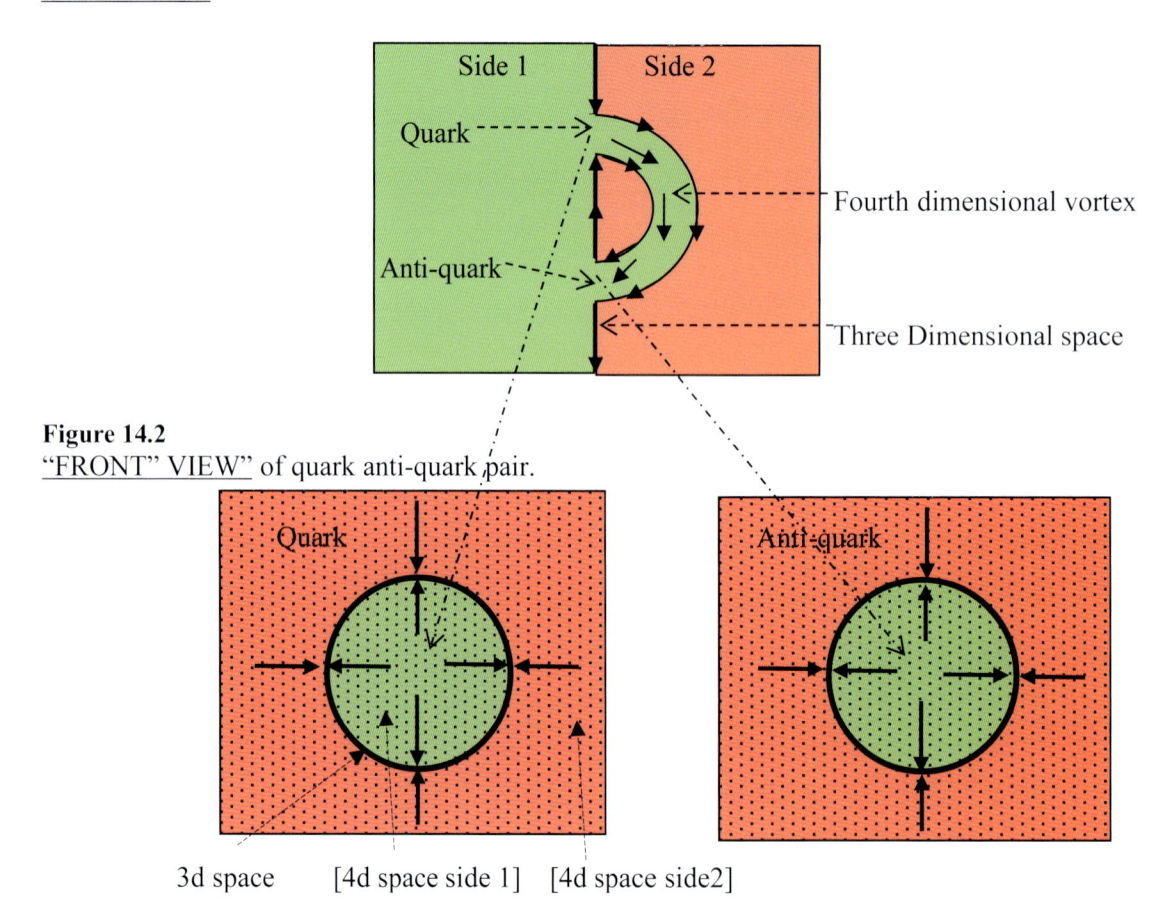

Figure 14.2
"FRONT" VIEW" of quark anti-quark pair.

3d space   [4d space side 1]   [4d space side2]

41

The quark is a hole in higher dimensional space that space flows into [represented by the dots in the circle]; the hole in higher dimensional space that space flows out of [represented by the dots in the circle]. In the above drawings, the dots represent the density of space.

Notice how the density of the space in the circular holes [from side 1] equals the density of the surrounding space [from side 2]. The arrows represent the force of the pressure. In the above drawings, they are all of equal length because the pressure of the outward push from the inside of the hole equals the pressure from the inward push on the outside of the hole.

However, if the size of the hole increases, the situation changes:

Figure 14.3

Equilibrium out of balance:

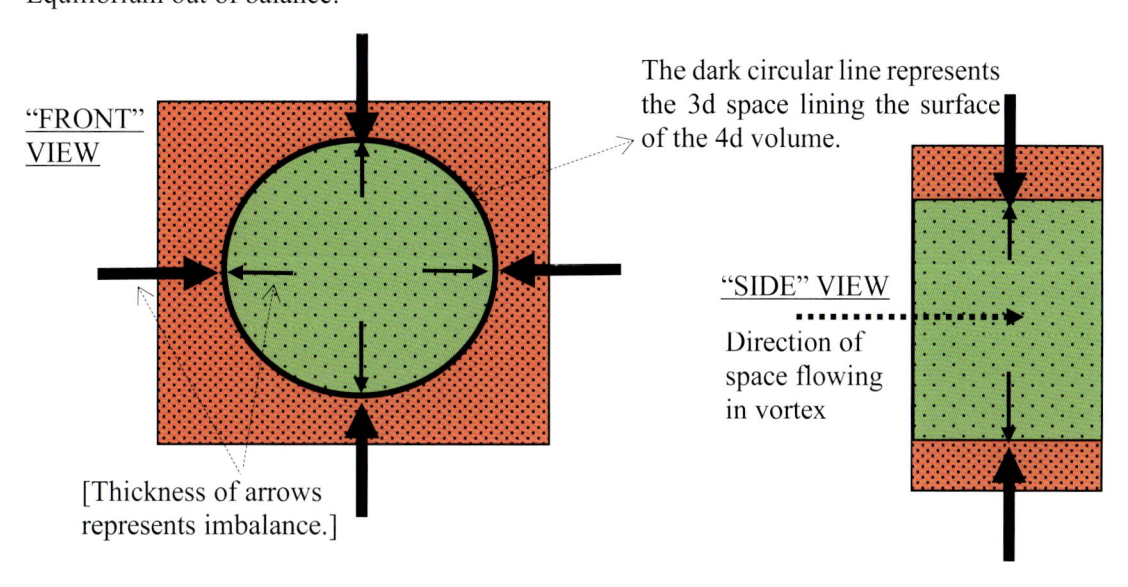

"FRONT" VIEW

The dark circular line represents the 3d space lining the surface of the 4d volume.

"SIDE" VIEW

Direction of space flowing in vortex

[Thickness of arrows represents imbalance.]

In the figure above, notice how the density of the space in the inside of the circle has decreased [*the dots are further apart*], and the density of the surrounding space has increased [*the dots are closer together*]. This situation destroys the equilibrium between the pressure of the two volumes of space as demonstrated by the size and the thickness of the arrows.

Because the two volumes of space are no longer in equilibrium, the outside pressure, [from side 2], causes the hole to collapse. It does so until the two pressures are again equal as seen below:

Figure 14.4

Equilibrium restored:

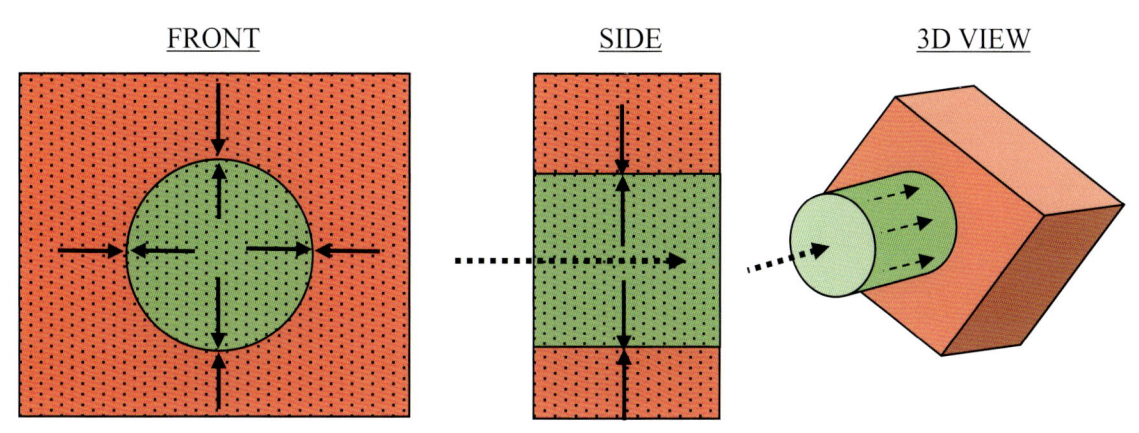

FRONT   SIDE   3D VIEW

> The Weak Force is <u>not</u> mediated by the W & Z particles. The W & Z particles are actually deflating vortices [such as when the vortex surrounding the neutron breaks from its torus shape and reassumes its tubular shape. This will be explained later in this book.
>
> The WEAK FORCE is actually <u>three forces in one</u>. And can now be defined as:
>
> The restoration of the equilibrium of the pressures of the two fourth dimensional sides of space [for 3d holes];
>
> And/or, the restoration of the equilibrium of the pressures of the two fifth dimensional sides of space [for 4d holes];
>
> And/or, the restoration of the equilibrium of the pressures of the two fourth dimensional sides of space [for 5d holes].

It should be noted here that in only two pairs - of the particles of nature - are the two sides of space in balance causing the Weak Force to be in a state of equilibrium: the proton, antiproton pair; and the electron, positron pair.

Chapter 15
Tunneling

> The Vortex Theory's discovery of vortices not only presents a new and revolutionary breakthrough about the workings of the subatomic world, it also reveals misconceptions about the subatomic world that need to be corrected. One of these is the mistaken belief that the electron can turn into a wave and pass through matter. Nothing could be further from the truth! The electron does not pass through matter at all. It passes "around matter" by traveling through fourth dimensional space. Here is how it happens.

THE TUNNELING ELECTRON

Without a doubt, strange things happen in the subatomic world. One of these is the well-known fact that the electron can seemingly move directly through matter.

For example, if we take a ball and throw it at a wall, the ball hits the wall and bounces back: but not so in the subatomic world. In the subatomic world, under the right circumstances, an electron shot at a wall of matter can seemingly pass right through it.

The electron reaches the wall, disappears, and then magically reappears on the other side. The 20th Century Science's explanation is just as amazing: and false!

Because the electron has both particle and wave characteristics, in the past, it was assumed by scientists that the electron is somehow turning into a wave, and then as a wave, somehow passing through matter in this way. However, with the discovery of the Vortex Theory, we know now that this is not true.

In Book 2, *The Vortex Theory*, it was explained how the particle characteristics of the electron are explained by its "three dimensional hole in space"; while the wave aspects of the electron are created by its denser region of surrounding space created by the outward flow of 3d space from the three dimensional hole.

Because the hole cannot exist without the denser region, and the denser region cannot exist without the presence of the hole, both are always present and dependent upon the other. Hence, the electron cannot somehow shed its "matter characteristics" [the hole] and pass through matter as a wave.

The key to understanding how the electron can pass through the wall is to realize that it is really just the end of a fourth dimensional vortex. The way the electron is able to pass through the wall of matter is to pull its end [the 3d hole] out of three dimensional space and move directly through 4d space until it reaches the other side. For example:

Figure 15.1

[a] THE ELECTRON APPROACHES A WALL OF MATTER:
[from our point of view]

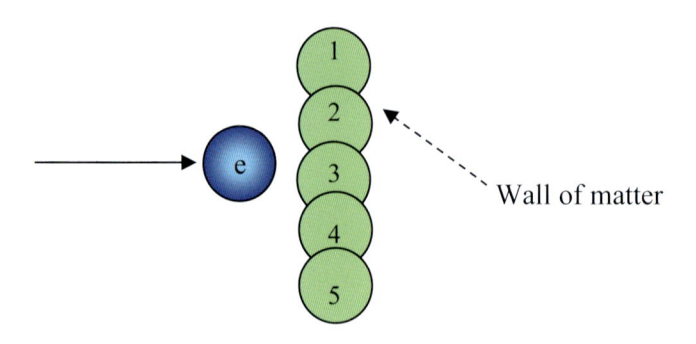

Wall of matter

[b] THE ELECTRON DISAPPEARS:

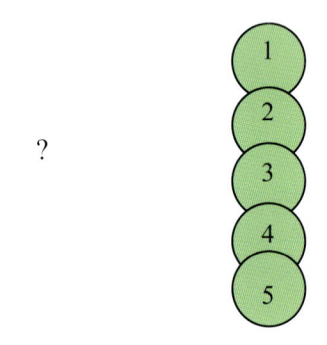

[c] LIKE MAGIC, THE ELECTRON REAPPEARS UPON THE OPPOSITE SIDE OF THE WALL:

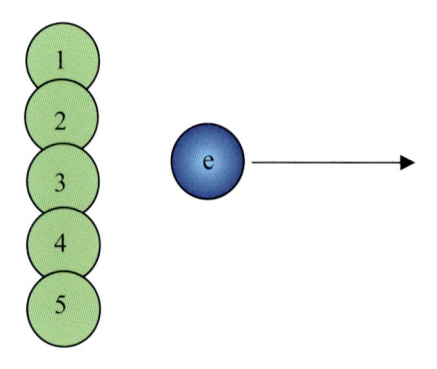

However, the above illustration was from <u>our point of view</u>. What really happens is seen below:

Figure 15.2

As seen below, the electron is really a 3d hole at the end of a vortex in 4d space: approaching a wall of matter [3d holes labeled # 1,2,3,4]:

3d space [black line]

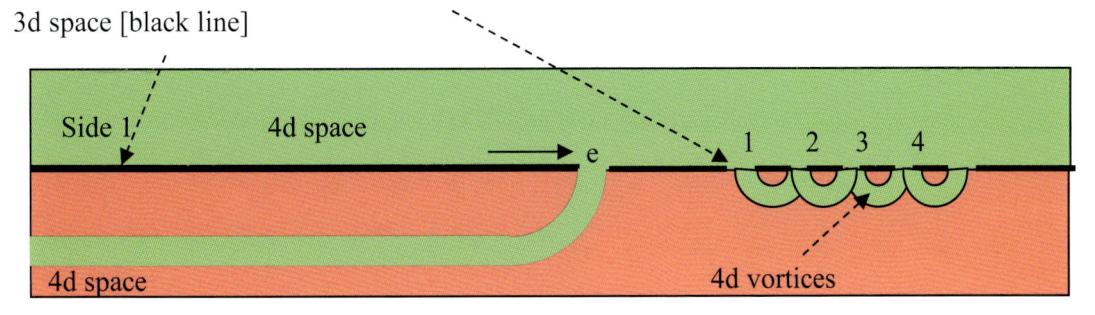

Figure 15.3

When the electron reaches the wall of matter [#1,2,3,4], its end dips down into 4d space and the vortex "Tunnels" through 4d space "under the wall".

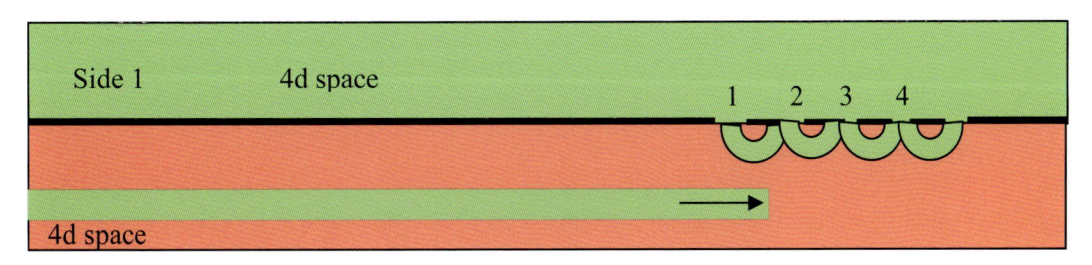

Figure 15.4

The electron [the end of the tunneling vortex] passes under the wall of matter and re-emerges on the other side. Jumping back upward into 3d space [black line], making it seem as if it had somehow miraculously passed through the wall.

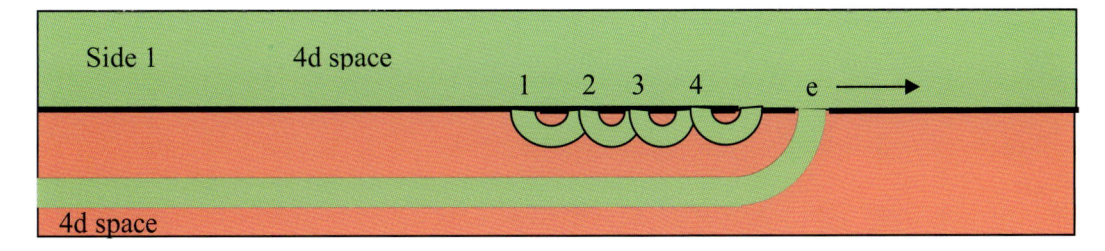

And now, the foundation is laid upon which to build the explanation of the neutrino:

Chapter 16
Neutrinos

It is perhaps an irony that the tiniest particle in the universe – the neutrino – is capable of explaining how the entire universe itself is constructed: that the universe is constructed out of one gigantic particle split into two parts – one in expansion, one in contraction; causing the one particle to turn inside out – reaffirming and giving further proof of the thesis presented at the start of this book.

NEUTRINOS

When the Vortex Theory was first discovered, one of its goals was to determine if it could explain how the neutrino was constructed. The neutrino has always appeared to be a difficult particle to understand because it seems to be a cross between matter and energy. Traveling at the speed of light makes it appear to possess the characteristics of energy, and yet, in its collisions with protons and neutrons it behaves as if it has mass like matter.

Initially, the first problem – the energy characteristic of the neutrino – seemed to be solved when it was discovered that such a particle could indeed exist if it was behaving like a transverse wave created upon the surface of fourth dimensional space. Just as a photon can be compared to a compression wave traveling through the volume of water within an ocean, a neutrino can be compared to a transverse wave traveling upon the ocean's surface. [With the exception being that both the photon and the neutrino are quantized.]

Using an analogy of a wave upon the surface of the ocean, an ocean wave travels upon the surface of the water like energy, but moves matter floating upon the surface as if it has mass. A similar situation would occur with the neutrino. Traveling upon the surface of space like a wave, it is similar to energy. The mass characteristic of the neutrino arises during collisions. When a neutrino strikes a particle of matter, the addition or subtraction of its volume of space in its interior of the particle makes it appear as if the neutrino possesses mass.

The anti-neutrino is bent outward, into side 2, and the neutrino is bent inward into side 1. The neutrino can be compared to a crest while the anti-neutrino can be compared to a trough. This relationship can be seen in the 2d to 3d analogy drawn below.

Figure 16.1

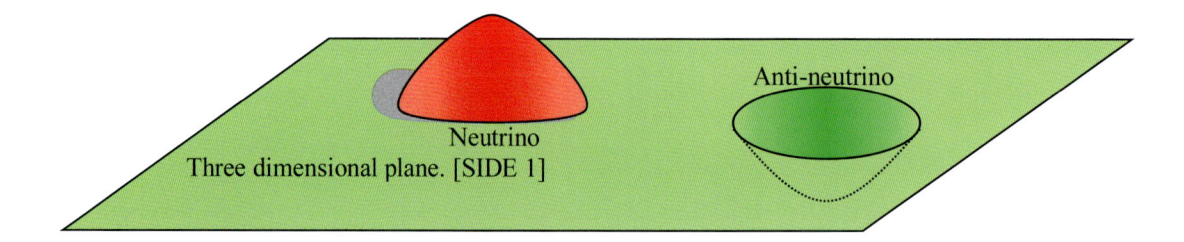

Figure 16.2

The neutrino and anti-neutrino as seen from Side 1 and Side 2:

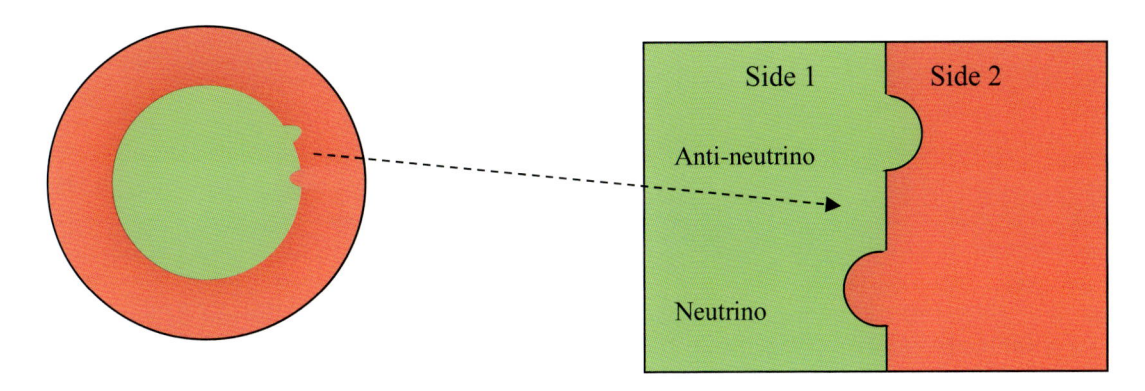

Note: how the neutrino is bent outward, into side 1, while the anti-neutrino is bent outward into side 2: the anti-neutrino is actually part of side 1, and the neutrino is part of side 2. This is a very important observation, because it reveals that there are two volumes of higher dimensional space. It also reveals that neutrino particles are unique, totally different from all other particles and waves.

As quantized waves, unlike the transverse waves of the ocean, they do not continuously spread out in all directions at once, but instead, maintain their spherical shapes. Also, these waves possess ½ spin. But unlike particles such as protons and electrons that also possess ½ spin, the ½ spin of neutrinos is created by the *spin of the surface they are created upon:* one volume spins CW, the other spins CCW. (This surface effect of the two different spins, when transferred to the neutrinos, creates the phenomenon that has come to be known as asymmetric parity, and will be explained shortly.)

However, before the great mystery called asymmetric parity can be explained, first, it needs to be explained that there are two ways of creating neutrinos: as deflations in space; and as waves created by "Tunneling particles."

In the figure below, an anti-neutrino can be created by the deflation of a particle. When the particle deflates on side 1, it pushes out-ward into side 2, creating the quantized wave seen below:

Figure 16.3

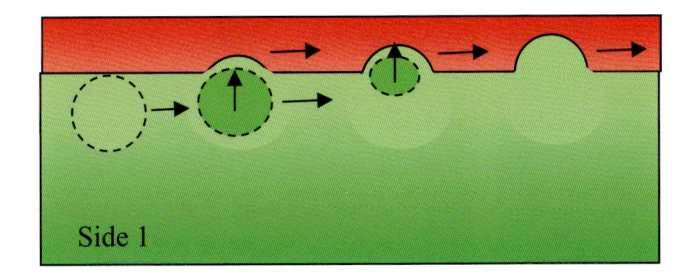

Chapter 17
Tunneling, the Creation of a New Class of Particle

> The initial discovery of the creation of neutrinos was made using the principles of classical physics. However, during an advanced study of neutrinos using the principles of the Vortex Theory, comes the fantastic discovery that these surface waves can also be created by the ends of vortices that are "tunneling" through space at the speed of light! The term "tunneling" was borrowed from the effect created by the electron in the tunneling diode.

Just as a submarine traveling close to the surface creates a surface wave, the end of a vortex not attached to the surface but instead traveling through a higher volume of space can also create a surface wave upon the lower surface.

Another fascinating characteristic of neutrinos is the fact that there are at present three different types: electron neutrinos and electron anti-neutrinos; muon neutrinos and muon anti-neutrinos; and tau neutrinos and tau anti-neutrinos. Later it was also discovered that just like the different layers of matter, each different type of neutrino is constructed on the surface of a different dimension of space.

Figure 17.1 THE TUNNELING ANTI-UP QUARK creates a [negative muon] neutrino

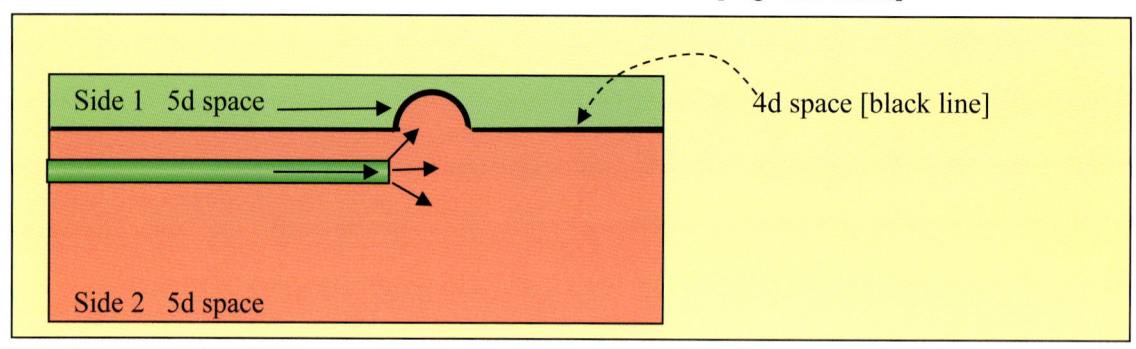

In the above drawing, an anti-up quark vortex whose length is expanding, is now tunneling through the 5d volume of side 2. Notice how the end of this expanding vortex from side 1 is no longer connected to the side 1 surface, but instead, travels exclusively through the volume of side 2; creating above it, a volume of space bent outward into side 1's 4d surface: creating a neutrino. Equally important, because the surface of 4d space, is actually a volume whose own surface is 3d space; this traveling transverse wave upon the surface of 4d space, also bends downward into *its* 3d surface, which just happens to the be the 3d volume of space we live in: [our 3d universe!]

Figure 17.2 4d neutrino also creates a 3d neutrino

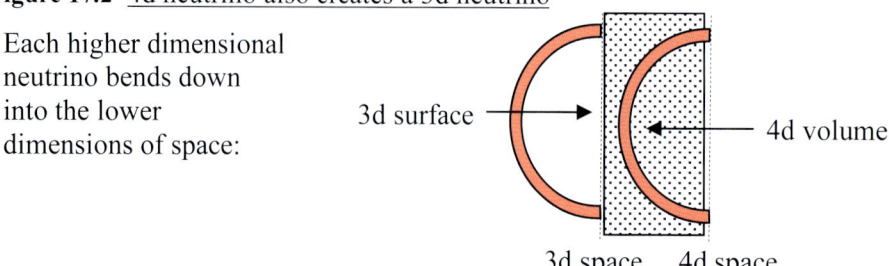

Each higher dimensional neutrino bends down into the lower dimensions of space:

[This is not conjecture but is explained later in Particle Collisions.]

Figure 17.3 THE TUNNELING UP QUARK creates an [positive muon] anti-neutrino

A <u>collapsing</u> vortex from side 1 can pull side 2 downward, creating a volume of space bent inward; creating an anti-neutrino.

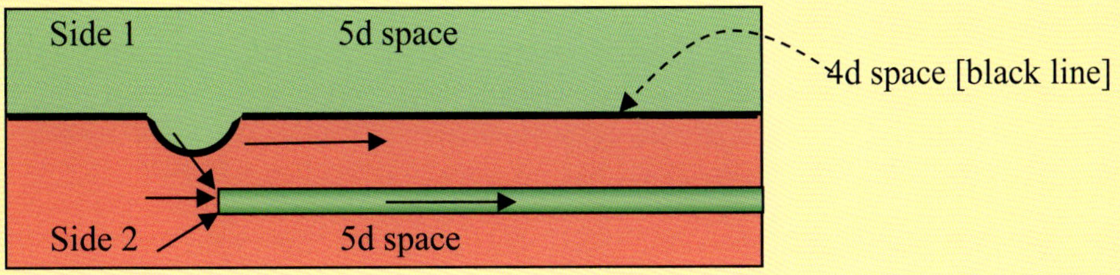

Side 1   5d space
4d space [black line]
Side 2   5d space

Figure 17.4 THE TUNNELING DOWN QUARK creates a [down] anti-neutrino

The creation of an anti-neutrino can also be created by a tunneling vortex whose end broke off from side 2 and is now traveling in side 1's volume. As the vortex moves through side 1's fifth dimensional volume, it pushes space outward, in front of it, causing an inward bulge into Side 2's surface: [This is also not conjecture but is explained in The Collision between a proton & an electron anti-neutrino in chapter 20]

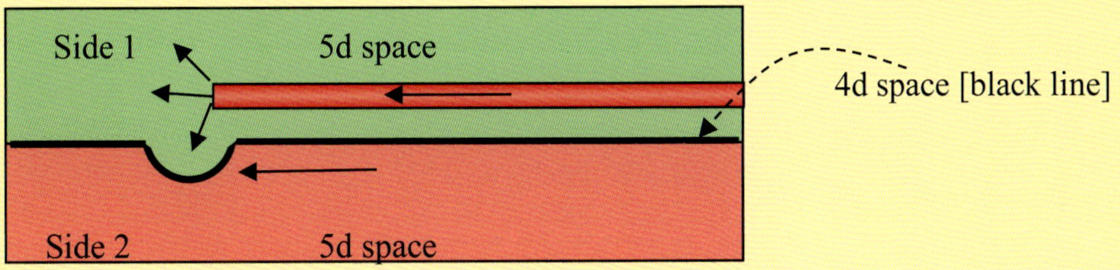

Side 1   5d space
4d space [black line]
Side 2   5d space

Figure 17.5 THE TUNNELING ANTI-DOWN QUARK

THE TUNNELING ANTI-DOWN QUARK creates a [anti-down] neutrino:
In the below drawing, notice how a <u>contracting</u> vortex from side 2 is pulling upward upon the surface of side 2, creating an upward bulge upon the surface of side 2, creating the neutrino:

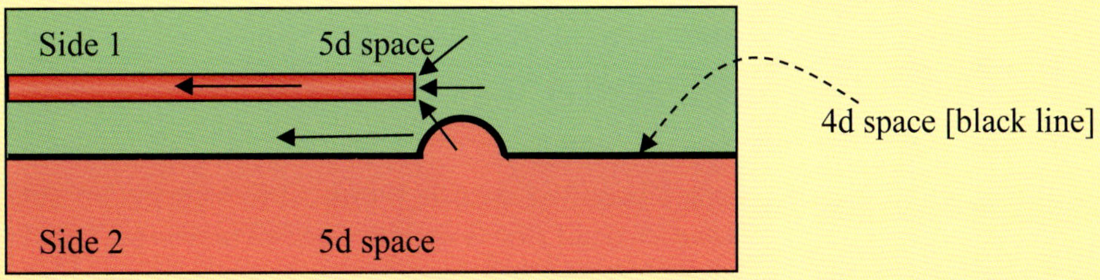

Side 1   5d space
4d space [black line]
Side 2   5d space

Figure 17.6 THE TUNNELING STRANGE QUARK creates an [strange] anti-neutrino

In the below drawing, notice how an <u>expanding</u> vortex from side 1 is pushing downward upon side 2, creating a downward bulge upon the surface of side 2, creating the anti-neutrino:

In the below drawing, notice how a <u>contracting</u> vortex from side 2 is pulling upward on side 1, creating an upward bulge upon the surface of side 2, creating the neutrino:

Figure 17.7 THE TUNNELING ANTI-STRANGE QUARK creates a [anti-strange] neutrino

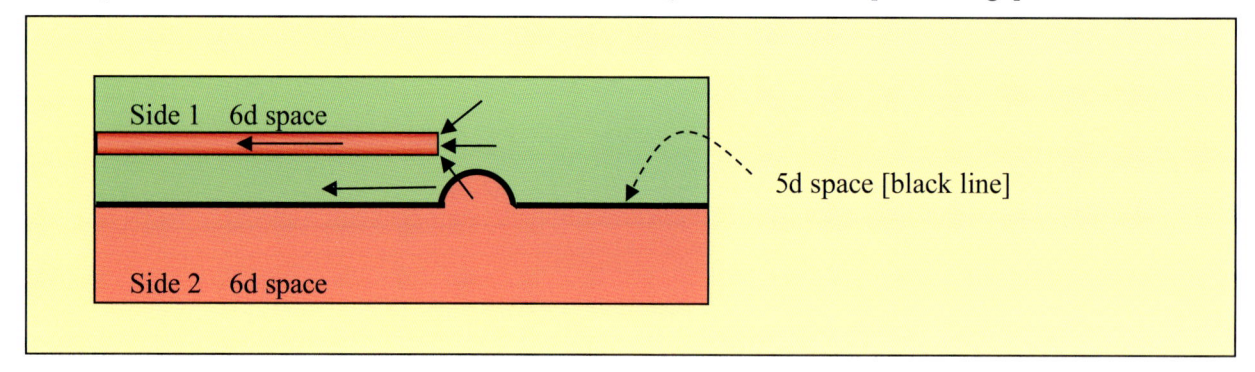

Figure 17.8 THE TUNNELING CHARM QUARK creates a [tau] neutrino

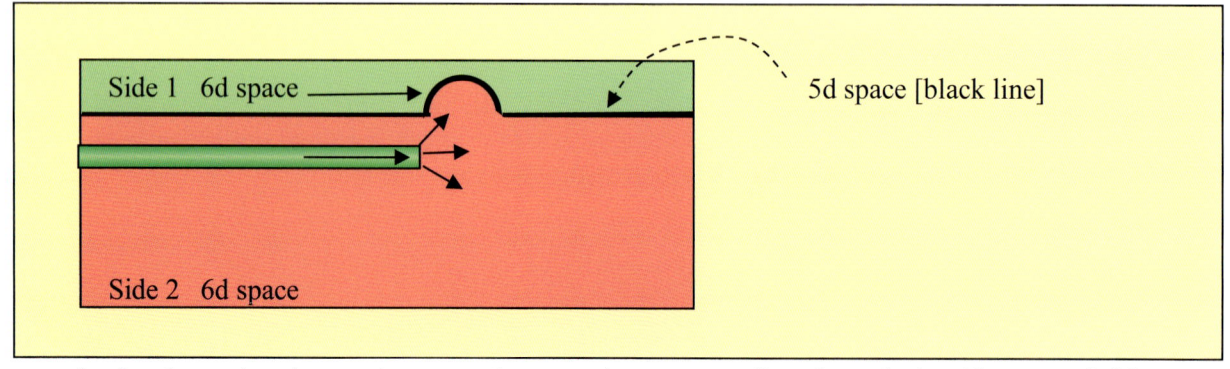

In the above drawing, a charm quark vortex is now tunneling through the 6d space of side 2. Notice how the end of this <u>expanding</u> vortex from side 1 is no longer connected to the side 1 surface, but instead, travels through the volume of side 2; creating above it, a volume of space bent outward into side 1's surface: creating a neutrino.

A <u>collapsing</u> vortex from side 1 can pull side 2 downward, creating a volume of space bent inward; creating an anti-neutrino as seen below:

Figure 17.9 THE TUNNELING ANTI-CHARM QUARK creates an [anti-tau] anti-neutrino

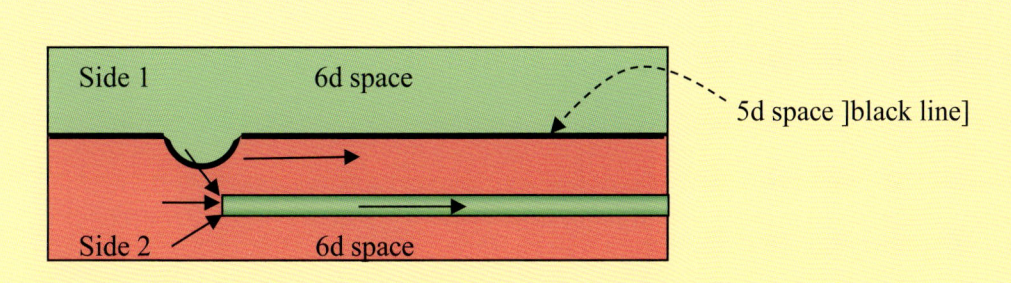

In the above drawing, an anti-charm quark vortex is now tunneling through the 6d space of side 2. Notice how the end of this <u>contracting</u> vortex from side 1 is no longer connected to the side 1 surface, but instead, travels through the volume of side 2; creating above it, a volume of space pulled inward into side 1's surface: creating an anti-neutrino.

Note: the TOP and BOTTOM QUARKS will also tunnel and create neutrinos and anti-neutrinos. However, these have not been seen yet because only the accelerator at CERN is capable of generating the energy to see them.

SUMMARY OF THE THREE TYPES OF NEUTRINOS:

Note, in drawing #1, notice how the electron neutrino and anti-neutrino exist only upon the 3d surface of 4d space; in drawing #2 notice how the muon neutrino and anti-neutrino existing upon the surface of 4d space also contain an electron neutrino and anti-neutrino component existing upon the surface of 3d space; but in drawing #3, the tau neutrino and anti-neutrino have components existing upon all three surfaces of higher dimensional space.

Figure 17.10

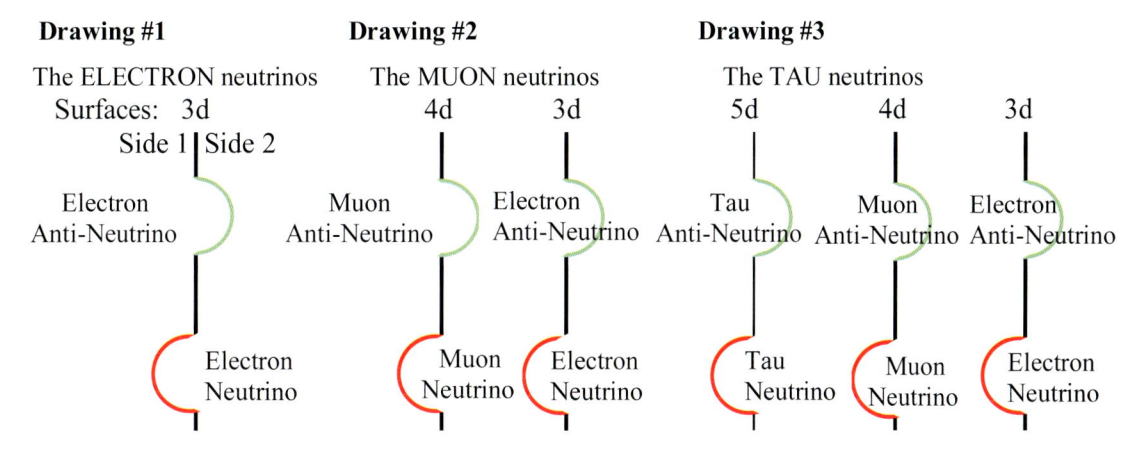

Drawing #1	Drawing #2	Drawing #3
The ELECTRON neutrinos	The MUON neutrinos	The TAU neutrinos
Surfaces: 3d	4d 3d	5d 4d 3d
Side 1 Side 2		
Electron Anti-Neutrino	Muon Anti-Neutrino / Electron Anti-Neutrino	Tau Anti-Neutrino / Muon Anti-Neutrino / Electron Anti-Neutrino
Electron Neutrino	Muon Neutrino / Electron Neutrino	Tau Neutrino / Muon Neutrino / Electron Neutrino

[The different tunneling components of the muon and tau neutrinos and anti-neutrinos will become important in their collisions between particles: especially the collision between the muon neutrino and the neutron.].

Chapter 18
Tunneling Particles!

> The term tunneling is associated with the silicon chips used in electronic circuits. Tunneling refers to a most curious action made by the electrons while in the electrical current flowing through the chip. When an electron is moving through the silicon chip and encounters the "junction barrier," it suddenly disappears, and then just as suddenly reappears upon the other side of the barrier.

As stated before, again, it is popularly believed that because matter behaves as both a particle and a wave; that when the electron encounters a barrier, the electron somehow transforms itself into a wave, travels through the barrier as a wave, and then on the other side, resumes its role as a particle and wave! This wave theory is given credence when it is realized that the electron travels at the speed of light through the barrier as a light wave would. However, this is a mistaken idea: a product of incorrect 20th Century scientific logic.

It is mistaken because the Vortex Theory reveals that the wave characteristics of matter are created by a region of dense or less dense space surrounding a particle. Because of this region of dense, less dense space, the idea that the electron can somehow be transformed into a wave is no longer valid. And suddenly, the explanation of tunneling is no longer valid either.

Instead, when the electron encounters the barrier in the tunneling diode, the three dimensional electron hole – that is nothing more than the end of the vortex of flowing space – leaves the surface and "tunnels" through 4d space. After it passes under the junction barrier, the vortex again penetrates the surface recreating the 3d hole we call the electron.

The key to understanding the misunderstanding about the electron traveling as a wave comes from the electron's speed through the barrier: the speed of light. Because an electron cannot travel at the speed of light, 20th Century theorists believed that it could only travel as a wave. However, this misunderstanding is cleared up when it is realized that the speed of the space flowing in the vortex flows at the speed of light. Consequently, when the end of the vortex [the electron] dips down into higher dimensional space, its open end tunnels through higher dimensional space at the speed of the lengthening vortex: the speed of light. And when the electron reappears upon the other side of the barrier, it seems as if the electron traveled at the speed of light.

In a similar manner, the ends of the vortices that quarks are made out of also tunnel through space. The first one we will discuss is the tunneling up quark.

THE TUNNELING UP QUARK

In Chapter 17, when looking at Figures 17.1 to 17.9, we see a representation of Quarks pushing into the higher dimensional space that spawned them and moving like phantoms at the speed of light through the invisible space made of something.

When it is upon the 4d surface, because space was flowing into the up quark, when it tunnels upon side 2 it is in fact deflating. Consequently, its vortex will continue to decrease in length until it reaches its other end – the anti-up quark where both will instantly disappear pulling side 1 outward creating a down, anti-down pair.

THE TUNNELING ANTI-DOWN QUARK

The anti-down quark tunnels are similar to the up quark except it tunnels on side 1. Its vortex also continues to deflate until it reaches the down quark where both will disappear, pulling side 2 outward creating the up, anti-up pair.

THE TUNNELING ANTI-UP QUARK

When the anti-up quark decays, because it is at the other end of a vortex that space is flowing into [the up quark], it continues on and on at the speed of light or until it impacts other matter.

THE TUNNELING DOWN QUARK

When the down quark decays, because its vortex is being fed by the anti-down, it continues on at the speed of light or until it impacts other matter.

Note: because up and down quarks have extremely long lifetimes within protons, it is assumed that they can survive decay, allowing their ends to tunnel. However, it is unknown at the time of this writing if the strange, charm, bottom, and top particles can also tunnel for long distances. Their increased mass might cause all of their ends to decay long before the other end can tunnel.

Chapter 19
The Asymmetric Parity of Neutrinos

> One of the great mysteries of the subatomic world is something called the "Asymmetric Parity of Neutrinos." A Nobel Prize was awarded for this phenomenon that has had no explanation – until now! The new vision of the universe presented at the start of this book is now able to explain what all of the mathematicians' of the 20th Century cannot explain.

THE ASYMMETRIC PARITY OF NEUTRINOS

When the asymmetric parity of neutrinos was first proposed by Tsung Dao Lee and Chen Ning-Yang in 1956, it was ill received – especially by Nobel Prize winner Wolfgang Pauli who was totally against it. The idea that all neutrinos appear to be "left-handed" [spinning counterclockwise] while all anti-neutrinos appear to be "right handed" [spinning clockwise] seemed to defy logic.

But then, in 1957, when Chien-Shiung Wu proved it was true, the world of physics was stunned, and Pauli had to apologize. However, even though it is more than fifty years later, this discovery still seems to defy all logic. Because why should neutrinos only spin counterclockwise and anti-neutrinos only spin clockwise when hadrons, mesons, and other particles – such as electrons and positrons – can spin either clockwise or counterclockwise? It doesn't seem logical!

The answer only seems illogical when scientists try to explain it using an incorrect vision of the universe. The answer, like all other answers, can be explained using the correct vision of the universe. In this situation, the correct logic is created by first classifying "particles" according to their construction.

In our mistaken vision of the universe, we group all subatomic structures together as "particles." Nothing could be further from the truth. Especially for neutrinos!

NEUTRINOS AS SURFACE WAVES

Mistakenly classified as leptons, unlike electrons, muons, and tauons, what we call neutrinos and anti-neutrinos are not "particles" of matter at all, instead, they are transverse waves created upon the surface of space. Just like particles can possess potential energy, neutrinos and anti-neutrinos can be said to possess "potential mass."

When neutrinos and anti-neutrinos collide with a "particle" of matter, their tiny amount of space either bent into or out of higher dimensional space is transferred into the volume or subtracted from the volume of the "particle" it hits. This transfer of volume gives it the characteristic of mass. However, neutrinos and anti-neutrinos are not holes in space. They are surface phenomena: quantized waves.

Although electrons and positrons appear to be surface phenomenon they are not. They are merely the three dimensional ends of fourth dimensional vortices of flowing space moving from side 1 into side 2 and back: flowing into and out of higher dimensional space. This is where they differ from neutrinos and anti-neutrinos. Unlike matter and anti-matter, neutrinos and anti-neutrinos have no vortices connecting them.

Because neutrinos and anti-neutrinos are transverse waves existing upon the surface of higher dimensional space – and *are created by the surface of the higher dimensional space they are formed upon* – their individual characteristics [such as their asymmetric parity] are determined by the surface characteristics of *higher dimensional space they exist upon.*

Figure 19.1

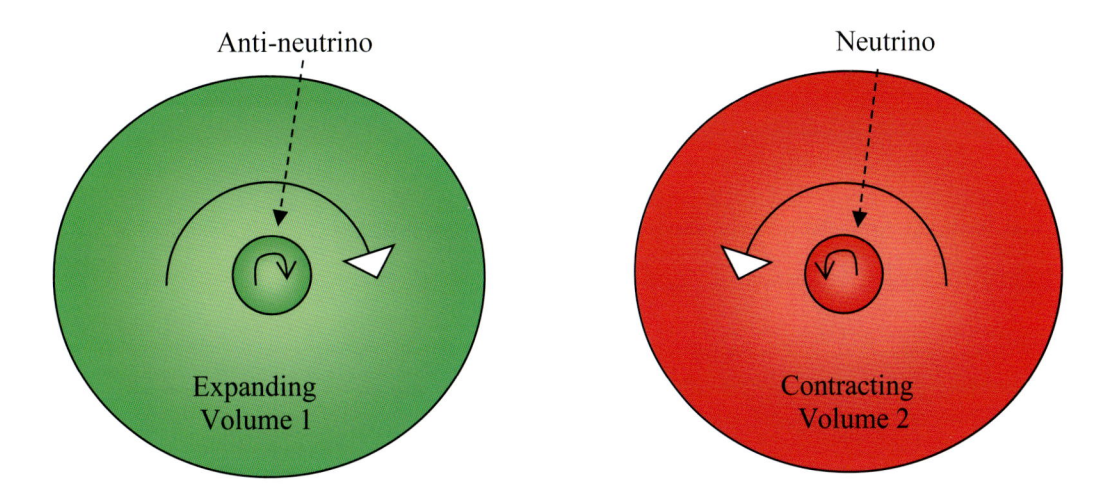

Anti-neutrino / Neutrino

Expanding Volume 1 / Contracting Volume 2

This is an important observation because it not only allows us to explain how neutrinos and anti-neutrinos are constructed it also gives us a glimpse into how the universe itself is constructed.

For example, if the universe was indeed created by a region of higher dimensional space turning inside-out creating two different volumes of space [one in expansion and one in contraction] then two characteristics would emerge that would be responsible for creating the asymmetric parity of neutrinos: #1 the rotating universe; and #2, the principle of "left becoming right" [to be explained shortly].

THE ROTATING UNIVERSE?

Recent studies conclude that the universe is not rotating. However, these scientists who made this statement, and who did this investigation do not know how the universe is constructed. If they did, they would realize that astronomical studies of Galaxies and photons will not reveal a rotation. This failure results from the fact that the universe is not constructed out of individual particles moving independently of each other. Instead, the universe is one single gigantic particle, and all matter is imbedded in its surface. Even light that seems to move independently on its own moves *within* the space of the universe.

Nor is there a Coriolis Acceleration. The Coriolis Acceleration of the earth is responsible for the rotation of hurricanes. However, it must be understood that the atmosphere of the earth is independent of the earth's surface. Hence it is not a part of the physical surface of the earth.

But the same is not true for the universe. There is nothing that moves independently of the space the universe is made out of. Everything is embedded in it. Hence, we will not see a circulation of galaxies responding to a Coriolis Acceleration of space; or a vast movement of photons in a circular pattern.

And yet, the rotation of neutrinos reveals that their two different and opposite rotations disclose that there are two volumes of space, each rotating in the opposite direction to each other. And this rotation is immense!

To get a better understanding of this perspective, take an outing on some clear night. Drive into the mountains or preferably the desert, far away from the light of the cities. Then once you've found a suitable parking area, step outside of your car, and look up at all the stars in the sky. Look hard and long at the stars and particularly the Milky Way, remembering as you do that this faint cloudy streak of light is really made up of hundreds of billions of stars; and these are only the ones we can see. Most are so faint we cannot see them because they are thousands of light years away; the Milky Way Galaxy is over a hundred thousand light years across, and the majority of the stars are only seen when looking through the most powerful telescopes.

Then, after having spent some time appreciating this great sight, turn on a flashlight, reach down and pick up a handful of sand. Next, let all of the sand fall away until you only hold a single grain in the palm of your hand. Then, as you look again at the stars, realize that everything you can see with the naked eye [except the Andromeda Galaxy] is rotating around one single fixed point – a point smaller than the size of the grain of sand you are holding in your hand! When you can look at the sky and visualize this rotation within your mind, it is a most profound experience!

When this experiment is properly done, it doesn't take a greater leap of faith to realize that the entire universe – like a spinning basketball – is spinning around a point smaller than this grain of sand!

Even harder to believe is the fact that when Ernst Mach was looking for his center of mass of the universe, he was looking in the **wrong place**. Unfortunately, the center of mass of the universe cannot be seen. Just like the center of rotation of a basketball is not upon its 2d surface but rather within its 3d volume, the center of rotation of the 3d space of our universe is not in three dimensional space. Instead, it is within a higher dimension. Because there are at least seven dimensions of space, it most likely exists within the volume of the seventh dimension [unless more dimensions exist].

WHEN LEFT BECOMES RIGHT AND RIGHT BECOMES LEFT!

The vision of space turning inside out is another principle possessing profound consequences.

If the physical universe was created by a region of space turning inside out and expanding into the volume it originally was a part of, then two volumes of space exist. Furthermore, everything in the two volumes is reversed:

To better understand how this reversal takes place, take an old white T-shirt, and use a black ink marker [that will bleed through to the other side] to draw a clockwise arrow on the left shoulder. Next, take off the shirt, turn it inside out; and then put it back on again.

When it is back on, look at both sleeves. Not only is the circular arrow on the inside of the shirt, it is now on the right shoulder, and the arrow is pointing counterclockwise!

And this is exactly the same situation with space.

For example, if the space of the universe originally existed as a "single particle" [as proposed in Chapter 3], then the substance it is made out of could only rotate as a single particle in one direction: say for example it is rotating clockwise.

However, if the center were turned inside out and began to expand outward into the existing volume [as also proposed in Chapter 3], its "left and right" [like the T-shirt] would become the opposite to the "left and right" of the space it is expanding into. This juxtaposition of left and right would create conditions upon one surface that would be directly opposite to those on the other surface: the expanding volume would now appear to be rotating counterclockwise; and objects ***attached*** to either surface will now appear to be rotating in opposite directions to each other.

So, if a portion of the expanding surface [Side 1] is pushed outward into the contracting volume of Side 2 creating an anti-neutrino, from the perspective of Side 2, this spherical hole will appear to take on the spin of its surface [Side 1]. It will seem to spin from left to right or clockwise from our point of view. Furthermore, if a portion of the contracting volume [Side 2] was pushed outward into the expanding volume of Side 1, creating the neutrino, from the perspective of Side 1, the neutrino will appear to take on the spin of Side 2. It will spin from right to left or counter-clockwise.

However, the same is true for the neutrino and the antineutrino: we witness the illusion and not the reality. We see the effect and not the cause. The neutrino and the anti-neutrino create the effect that they are in rotation because the surface of the space they are created upon is slowly rotating. Their rotations also appear to be opposite to each other because the two different surfaces of space they are created upon are in opposite rotation to each other.

Note: it must be mentioned that even though "particles" such as barons and leptons are holes upon the surface of 3d space. The 3d volume that flows into them, [or out of them], can either twist clockwise or counterclockwise as it exits into 4d space because it flows into and out of the vortices that are no longer part of the three dimensional surface. [Note too: photons are created *within* the 3d volume and not *upon* the 3d surface like neutrinos.]

Chapter 20
Charged Leptons: [the Electron, Muon, & Tau]

> The 20th Century vision of the universe that states charged leptons have no quarks within them is another mistake. In this section we will seek to disprove this false assumption. For if the more massive leptons are indeed holes within holes, then the higher dimensional holes they exist within, is either a new form of quark, or an adoption of preexisting quarks.
>
> Also, the hierarchy of holes and the number of dimensions present indicate that **there is at least one more charged lepton with its accompanying neutrino that has not yet been discovered: "the Wow!"**

LEPTONS

Although there are many mysteries in particle physics, the one that is the most intriguing is the mystery of the "three" leptons: the electron, muon, and tau? Why only these three? Why aren't there any intermediary particles in-between them?

The mystery is finally explained when it is realized that like quarks, the electron, muon, and tau leptons are created in "layers." These layers are created by the surfaces and the volumes of the different dimensions of space. The electron is a three dimensional hole existing upon the surface of the fourth dimension of space; the muon is a fourth dimensional hole existing upon the surface of the fifth dimension of space; and the tau is a fifth dimensional hole existing upon the surface of the sixth dimension of space.

Electrons and positrons are the fundamental particles that all other leptons are sheathed within: the neg. muon exists within the electron; and the neg. tau exists within the neg. muon, creating three layers of matter.

[Note: the electron positron pair are also the fundamental 3d holes that all other particles of matter possessing quarks exist within: such as, baryons, pions kaons, etc.]

The Electron

The simple construction of the electron is shown below. The electron is a three dimensional hole in space. Within this three dimensional hole in space is a fourth dimensional volume.

Figure 20.1

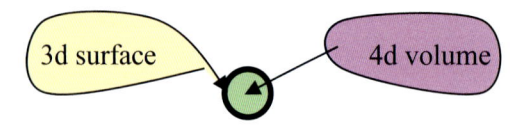

The electron possesses a definite size. However, when another hole is created within the fourth dimensional space inside its interior, the electron's size expands creating the muon.

The Muon

The construction of the muon is drawn below. Note that the muon is a fourth dimensional hole existing within the expanded three dimensional hole called the electron.

Figure 20.2

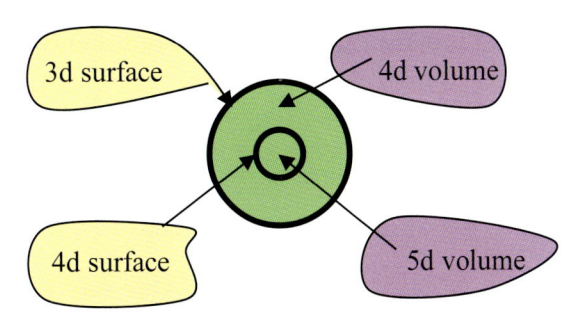

The fourth dimensional hole within the muon also possesses a definite size. However, when a hole is created within its fifth dimensional volume, both its size and the size of the electron expand to create the Tau.

The Tau

The construction of the Tau is drawn below. Note how the tau is actually constructed out of three layers of matter. The tau is a fifth dimensional hole existing within a fourth dimensional hole (the muon), existing within a three dimensional hole, (the electron).

Figure 20.3

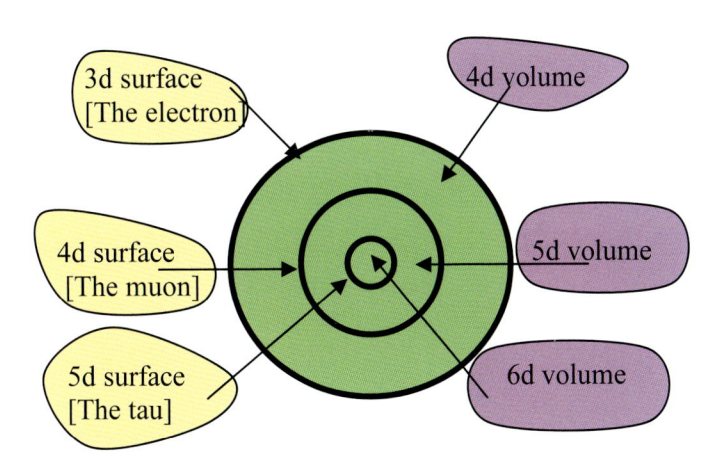

IMPLICATIONS OF THE 4TH AND 5TH DIMENSIONAL HOLES WITHIN LEPTONS!

The Vortex Theory's analysis of subatomic particles has discovered many revolutionary principles. These revolutionary principles have necessitated the re-evaluation of some widely accepted but erroneous beliefs. One of these is the absence of quark content within the muon and tau leptons.

When the analysis of leptons was conducted using the principles of the Vortex Theory, there appeared to be no reason to challenge the contemporary scientific belief that leptons contained no quarks. But after it became obvious that the differences in the masses of the electron, the muon and the tau could be explained by the presence of fourth and fifth dimensional holes, the preliminary hypothesis was inescapable: *the fourth dimensional hole within the muon was a quark; and the fifth dimensional hole within the tau was also a quark!*

The words "preliminary hypothesis" were used because the idea of quarks within muons and tau's seems to create as many problems as they answered. However, while using the principles of the Vortex Theory to analyze the decay of the muon and the tau, the evidence soon became overwhelming. Although, in the section entitled "Particle Decays," the explanation will be given on both how and why quarks decay or "Change Flavor," some of the results must be given here regarding the decay of the charm quark:

In studying the various decays of the tau particle, particle physicists have discovered that even though the tau is a lepton that supposedly possesses no quarks, there is the possibility of it decaying into particles containing quarks. Even more interesting is the fact that the tau is decaying into no quark more massive than the strange quark[1]. This is an observation of enormous importance, because using the new and revolutionary principles of the "2 sides of space" to explain particle decays, the Vortex Theory reveals that the tau is decaying similar to a charm quark. This observation forces us to consider the possibility that the fifth dimensional hole within the tau could either be a charm quark or a new quark similar to the charm.

Furthermore, because the charm quark is really a 5d hole existing within a 4d hole – [an up quark] – if the tau is created by the presence of a charm quark, then the muon is created by the presence of an up quark. And this is where a problem arises.

The up quark possesses a charge of only two thirds the value of the electron. Because the charge on the electron represents a volume of space flowing into a 3d hole, how can this volume then continue its journey onward into a 4d hole – capable of accepting only two thirds the volume of the flow coming out of the electron? This does not seem possible.

In our everyday experiences in three dimensional space, if we try to pour nine liters of water a second into a pipe that will only allow the passage of six liters a second, three liters don't make it into the pipe. This excess water either backs up and flows out onto the ground, or if the system is enclosed, the rate of flow slows down; decreasing the velocity of the flow within the pipe, [decreasing its "charge"], an effect that is not seen in the subatomic world.

In the subatomic world, if the velocity of space flowing into and out of particles changed, their electrostatic "charges" would be variables instead of constants. Since this is not seen, the volume of space flowing into and out of these particles is a constant. Also, if the excess one third volume was somehow flowing around and past the 4d hole, it would have to return to the surface of 3d space somewhere creating an exotic electron like particle with one third its charge. Because no such particle has ever been seen, there must be some other explanation not yet deduced that is responsible for allowing a quark with two thirds charge to exist alone within a particle containing one full charge: a seemingly impossible effect!

Is there an explanation?

THE EXPLANATION

In our daily life, we do not encounter "holes within holes." Because we don't, when we first try to imagine such a bizarre physical concept, we try desperately to conjure up mental images from

---

[1] Lide, David R, CRC Handbook of Chemistry and Physics, 2004, Lepton Summary Table, page 11:4

our everyday experiences and find that there are only analogous similarities, and this is where a problem is created.

To us, all holes are the same. In 3d space, holes are all two dimensional. They only possess length and width. Consequently, the only characteristics that differentiate one hole from another are their sizes and shapes. But this is not true for higher dimensional holes.

When we encounter higher dimensional holes, we stumble across one of the most shocking and disturbing twists of reality ever encountered: *the surface area of the higher dimensional hole is larger than the surface area of the lower dimensional hole encircling it.*

This seemingly impossible relationship can be expressed mathematically. For example, examining the mathematical relationship between a three dimensional hole and a fourth dimensional hole, we find that the surface area of a three dimensional hole is $S = 4\pi r^2$, while the surface area of a fourth dimensional hole is $S = 2\pi^2 r^3$. If the radius r in the three dimensional hole was equal to a value of 1.00m, and the radius r in the fourth dimensional hole was also 1.00m, when the values are solved for, the surface area of the 3d hole is $12.57m^2$, and the surface area of the 4d hole is $19.74m^3$. And it is easy to see that the surface area of the fourth dimensional hole is larger than the surface area of the three dimensional hole it is encased within. With one most notable exception!

When we look at the units of the surface area, we see that the 4d hole also possesses one more dimension $[m^3]$ than the 3d hole $[m^2]$. This is the most important difference because it tells us that the space flowing into the 3d hole $[m^3]$ is not the same as the space flowing into the 4d hole $[m^4]$: *it is constructed differently!* Since 3d space is flowing into the 2d surface of the 3d hole, and since the interior of the 3d hole is the surface of 4d space, it is *the 4d volume of space within the 3d hole [the electron] that is flowing into the 4d hole [the quark].*

For example: [from *The Vortex Theory*]

Because the fourth dimensional hole trapped within a three dimensional hole is impossible to draw, a two dimensional to three dimensional representation can be used.

In the figure below, notice how a two dimensional hole on the two dimensional plane has the two dimensional surface flowing into it:

Figure 20.4

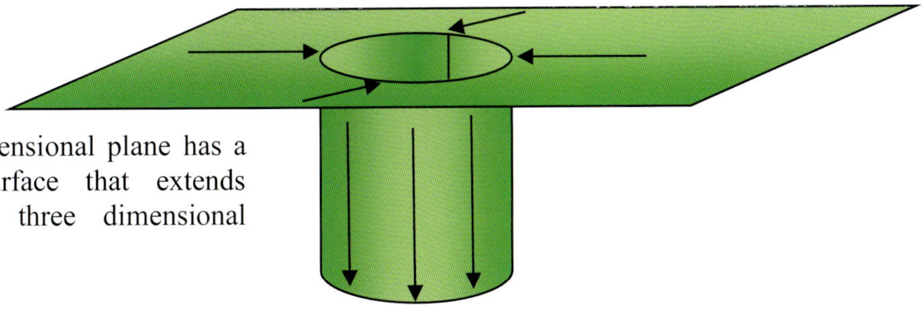

Here, a two dimensional plane has a hole in its surface that extends downward into three dimensional space.

Figure 20.5

Top view:

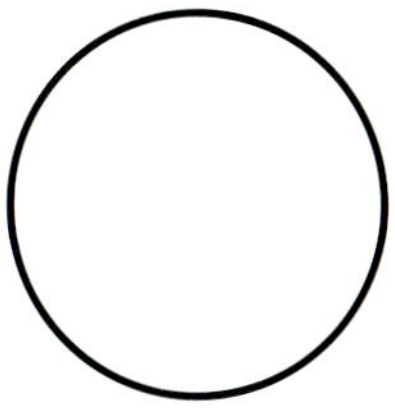

Figure 20.6

Side view:

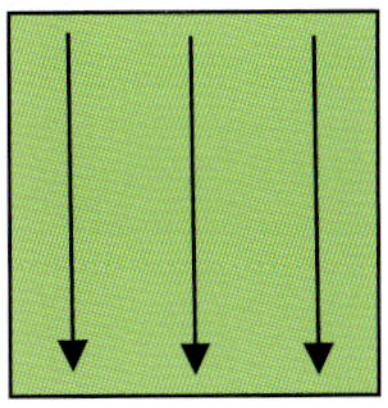

But now, imagine that a three dimensional hole exists within the two dimensional hole:

Figure 20.7

Here, a two dimensional plane has a 3d hole within the 2d hole upon its 2d surface:

Figure 20.8

Top view

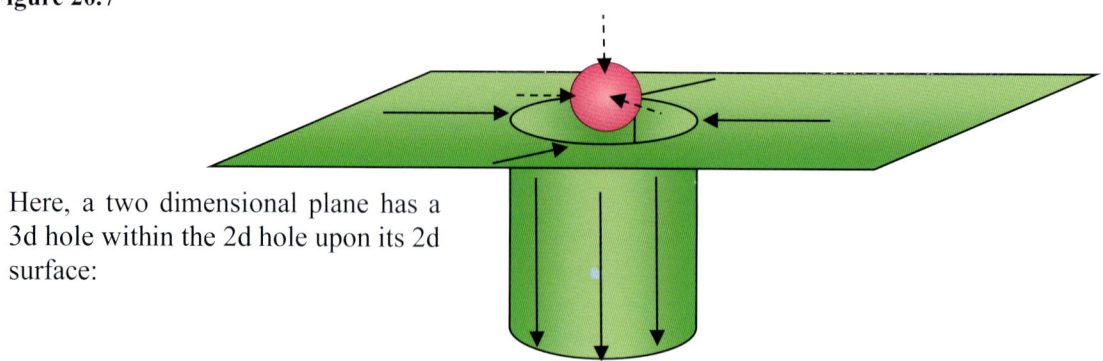

Figure 20.9

Side view:

Notice that none of the 2d space flowing into the 2d hole flows into the 3d hole it encircles! So, what is the relationship between the space flowing into the 2d hole and the space flowing into the 3d hole? *Answer: there is none*!

The only relationship between the 2d hole and the 3d hole is their size relationship. Only a certain size 2d hole can accommodate a certain sized 3d hole within it.

Transferring this concept to the 3d space to 4d space relationship in charged Leptons, only a certain sized 3d hole can accommodate a certain sized 4d hole within it.

So, how does the [m^2] surface and the [m^3] surface relate to each other if one is not flowing into the other? And is there a way to make a +1 charge seemingly flow into a +2/3 charged "particle"? Yes: by changing the density of space.

Because 4d space is flowing into the 4d hole, a fourth dimensional volume of space has to be present. But if the radius of the 3d hole and the 4d hole are the same [1.00m] then, the 3d hole is directly against the 4d hole creating a problem. At the points where the 3d hole touches the 4d hole, no fourth dimensional space can flow into the 4d hole. Consequently, either the 3d hole has to increase in size to allow 4d space to flow into the 4d hole, or the 4d hole has to decrease in size.

To determine which hole increases or decreases we examine the muon and find that its diameter is larger than the diameter of the electron. Hence, the electron has to increase in diameter to change into the muon. The amount the electron must increase is based upon two factors: #1, the charge on the muon, and #2, the density of space surrounding it. And this is where a major discrepancy is encountered...

The +2/3 charge on the up quark is misleading. As, stated previously, because of the difference in the densities of the space existing upon SIDE 1 and SIDE 2, the volume of space flowing into the up quark is double the volume that flows into the down quark - making the charge on the up quark twice the value of the charge on the down quark. If we had nothing else to reference these charges to, we would have to designate the up charge as 1 and the down charge as ½, or the up charge as 2 and the down charge as 1. So, when these two charges are referenced to the +1 charge on the particle they exist within, they merely appear to exist in increments of thirds. Consequently, the 2/3 charge on the up quarks and the 1/3 charge on the down quarks are relative values. Quarks only appear to possess these charges from our point of reference as seen here in 3d space.

When a single quark such as the up quark is present within the –1 muon, there is no other quark to reference it to. The only reference charge present is the charge on the muon which is –1. So, when the diameter of the electron expands to allow 4d space to flow into the up quark, and since the surface area of the up quark is actually larger than the surface area of the electron, the electron expands until it encircles a volume of 4d space that all flows into the up quark. Consequently, from our point of view in 3d space because the charge of the muon is –1, the charge on the up quark now appears to have changed to a value of –1!

[Note: if the surface area of the up quark was smaller than the surface area of the electron, the volume of 4d space flowing into the up quark could never be equalized by the expansion of the electron. There would always be an excess 4d volume of flowing space (a smaller 4d vortex) left over whose 3d surface would flow back into 3d space creating another particle possessing a partial charge.]

The implications of density changes in space is better understood by recalling the discovery in *The Vortex Theory* entitled - WHY ALL PARTICLES POSSES THE SAME AMOUNT OF CHARGE":

"Because of the inverse relationship between the size of the hole and the density of the space surrounding it, all particles [all three dimensional holes] possess the same amount of charge. *[The larger the hole the less dense the space surrounding it - making a smaller volume flow into a larger hole; and the smaller the hole the more dense the space surrounding it – allowing a larger volume to flow into a smaller hole.]*

This exact same relationship between the density of space and the volume flowing into the hole occurs in higher dimensional space. Consequently, a large enough volume of space has to be present to allow a dense or less dense region of space to form around the hole. If a hole was formed in a region of space where the less dense or denser region was restricted in size, the ability of space to flow into the hole would be restricted. Such a restriction would change the volume of the flow - changing the "charge" of the particle. This is precisely why the diameter of the electron increases.

When a fourth dimensional hole is created within the positron, fourth dimensional space begins to flow into the hole. This inward flow pulls additional fourth dimensional space into the hole, expanding the volume of the fourth dimensional space within the electron, expanding the size of the electron [the electron is the surface of this hole]. The electron continues to expand in size until a point of equilibrium is reached where all of the fourth dimensional space within the electron is flowing into the up quark. This increased size of the electron combined with the presence of the up quark creates the muon.

It should be noted that although the electron is a three dimensional hole *encircling* a region of fourth dimensional space, it does not enclose its fourth dimensional access. [A wall encircling a fortress restricts only two dimensional access from the ground, it does not limit three dimensional access from above.] Hence, all of the fourth dimensional space within the hole is not suddenly pulled into the up quark halting the flow. Instead, 4d space is able to continuously flow into the fourth dimensional hole keeping it open. [space becomes less dense increasing the surface area of the up quark allowing more space to flow into it, increasing its charge from 2/3 to 1.]

This variation in the diameter of the three dimensional hole allowing a larger or smaller volume of fourth dimensional space to flow into the fourth dimensional hole is equivalent to a "valve." This valve is the Mechanism that allows the + muon to change the charge on the up quark with a +2/3 charge to a +1 charge. But it also must be remembered that the charge on the muon and the charge on the quark are created by two completely different types of flowing space.

Within the expanded electron, these extra volumes of fourth dimensional space can be expressed using this schematic drawing. Although these volumes are not proportional, they nevertheless represent three regions of fourth dimensional space that will enable us to understand how and why neutrinos are created:

Figure 20.10

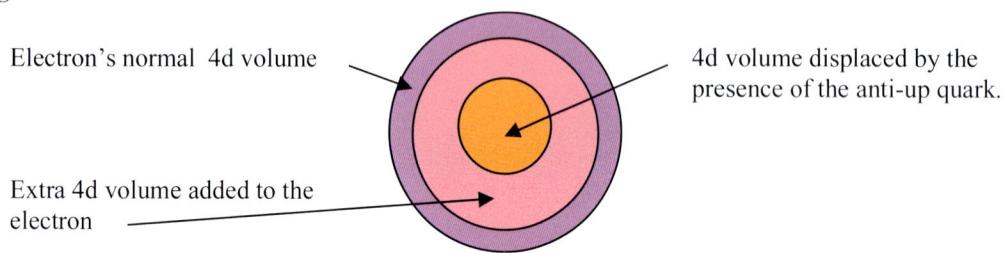

Electron's normal 4d volume

4d volume displaced by the presence of the anti-up quark.

Extra 4d volume added to the electron

This variation in the diameter of the three dimensional hole is the adjustment that allows what is perceived from our point of view to be a "anti-up quark" with -2/3 charge, to increase in value to -1, allowing it to exist within a negative muon possessing a charge of -1, [or +2/3 and +1 for an up inside a positive muon].

This exact same relationship occurs between the fourth dimensional hole within the muon and the fifth dimensional hole within the tau.

Figure 20.11

This figure represents the interior of the fifth dimensional hole within the up or anti-up quark:

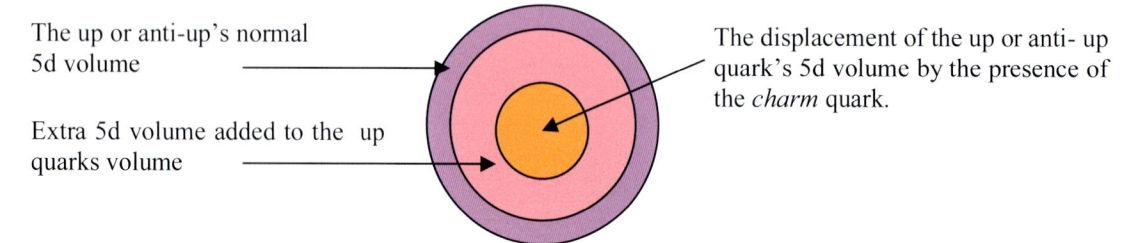

The up or anti-up's normal 5d volume

The displacement of the up or anti- up quark's 5d volume by the presence of the *charm* quark.

Extra 5d volume added to the up quarks volume

The effects of these extra volumes of space in the creation of neutrinos will be examined later in the decays of the tau and muon.

[Note, even though from our point of view the up quark appears to possess a +2/3 charge, from the point of view of the 5d hole, this value is irrelevant. From the point of view of the 5d hole, it exists alone within one 4d hole. Consequently, the space flowing into the 4d hole represents one volume. So, from its perspective, the charge on the 4d hole is not +2/3rds, but rather, represents a value of +1 (one volume)].

Note too: because the charm quark is a hole existing within the up quark, the charge on the charm quark cannot be measured. It is hidden from view. Only the charge on the up quark can ever be "measured."

A FOURTH LEPTON?

Once upon a time, a scientist wrote a paper stating that there could only be three charged leptons. Unfortunately, he did not know of the Vortex Theory, nor did he know how space is created; nor how many dimensions there are in the universe! Unfortunately, he believed in the Standard Model!

According to the *Standard Model's vision* of the universe everything is constructed out of "particles." Furthermore, the number of quark and lepton particles is complete. There are three "layers of Quark particles": the up, the down; the charm the strange; and the top and bottom quarks. There are three "layers of charged Lepton particles": the electron; the muon; and the tau. And there are three "layers of neutrino particles": the electron neutrino; the muon neutrino; and the tau neutrino. However, according to the *Vortex Theory*, not only is the Standard Model's vision of the universe a mistake, it is incomplete. There has to be a fourth charged lepton and a fourth neutrino.

According to the Vortex Theory, quarks are holes existing within holes and leptons are also holes existing within holes. And when we compare the number of holes we call Quarks to the number of holes we call Leptons, we come up one short. Observe the following in the figure below:

The first "layer of matter" starts out with a three dimensional surface hole [we call the **electron**] in the fourth dimension volume of space. Notice that the green 3d volume is actually the surface of the red 4d volume:

Figure 20.12

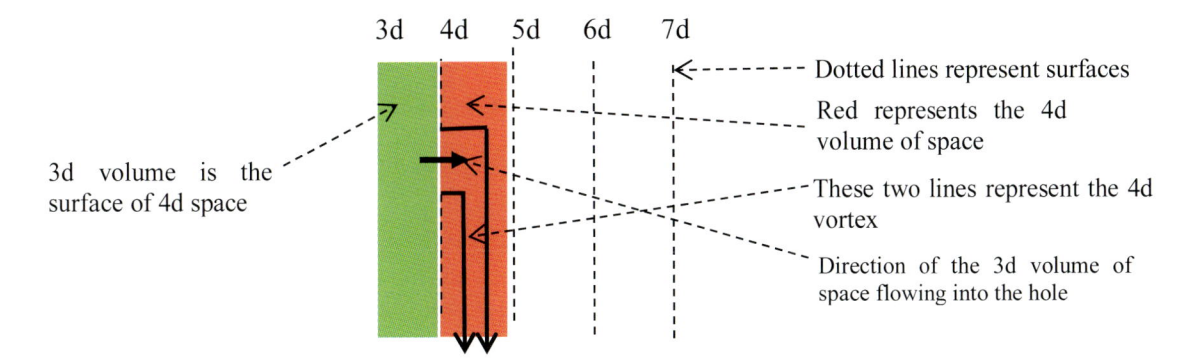

The second "layer of matter" starts out with a fourth dimensional surface hole [we call the **up quark**] in the fifth dimension volume of space. Notice that the red 4d volume is actually the surface of the violet 5d volume:

Figure 20.13

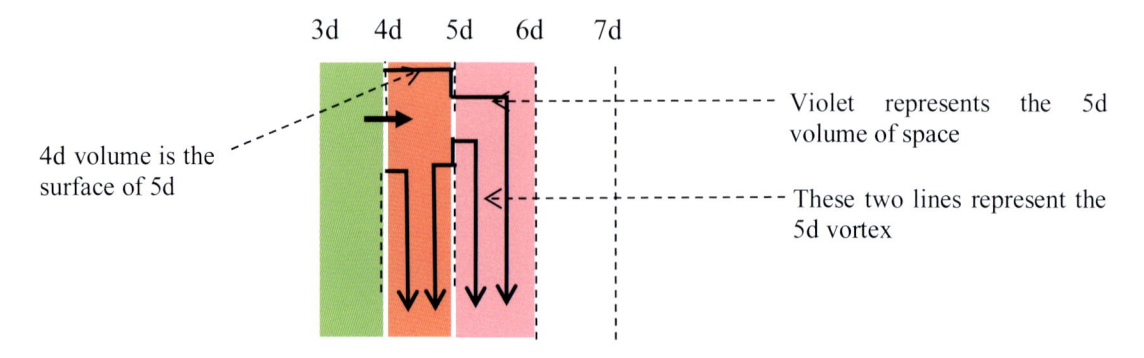

4d volume is the surface of 5d

Violet represents the 5d volume of space

These two lines represent the 5d vortex

The third "layer of matter" starts out with a fifth dimensional surface hole [we call the **charm quark**] in the sixth dimension volume of space. Notice that the violet 5d volume is actually the surface of the turquoise 6d volume:

Figure 20.14

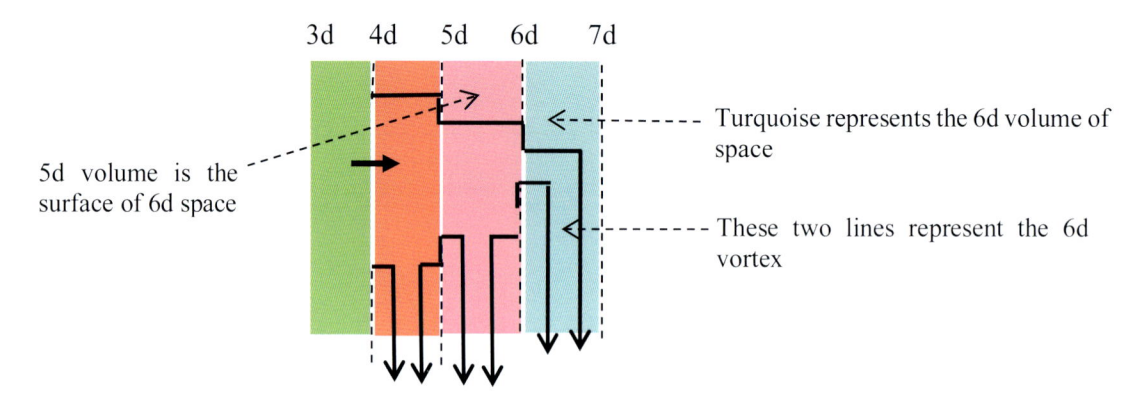

5d volume is the surface of 6d space

Turquoise represents the 6d volume of space

These two lines represent the 6d vortex

The fourth "layer of matter" starts out with a sixth dimensional surface hole [we call the **top quark**] in the seventh dimension volume of space. Notice that the turquoise 6d volume is actually the surface of the blue 7d volume:

Figure 20.15

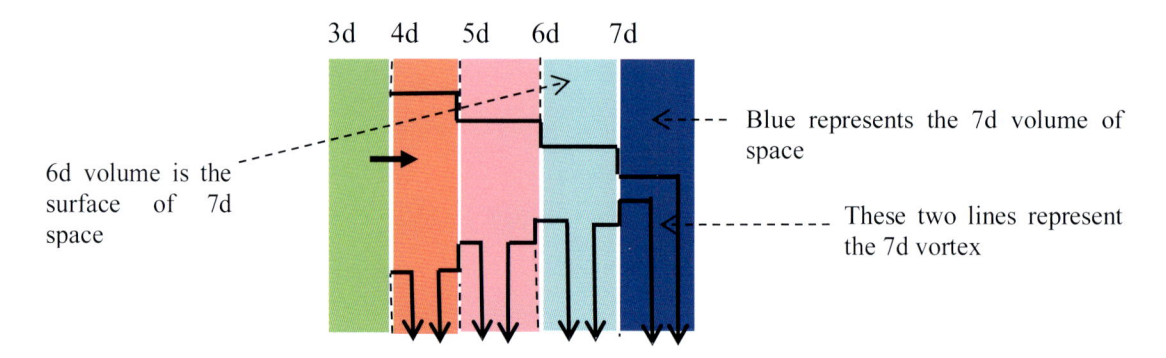

6d volume is the surface of 7d space

Blue represents the 7d volume of space

These two lines represent the 7d vortex

67

When we come to charged leptons, we observe a similar situation of holes within holes: The first "layer of matter" starts out with a three dimensional surface hole [we call the **electron**] in the fourth dimension volume of space. Notice that the green 3d volume is actually the surface of the red 4d volume:

Figure 20.16

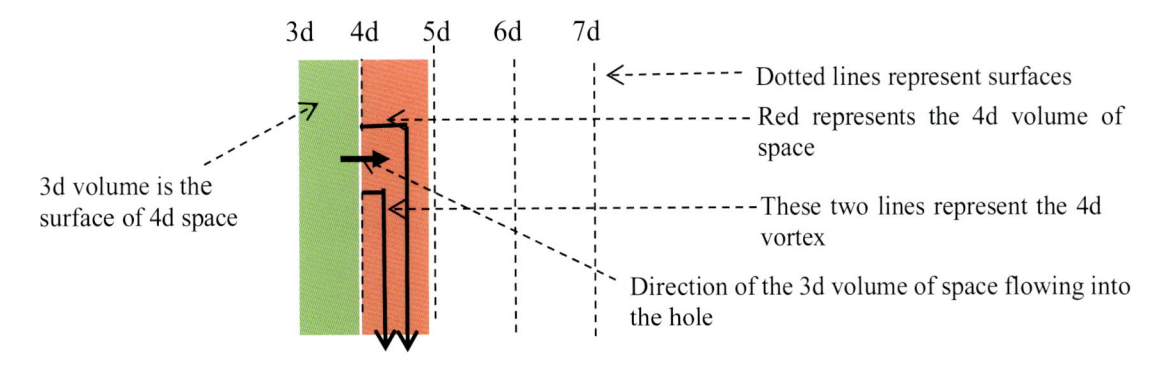

The second "layer of matter" starts out with a fourth dimensional surface hole [we call the **muon**] in the fifth dimension volume of space. Notice that the red 4d volume is actually the surface of the violet 5d volume:

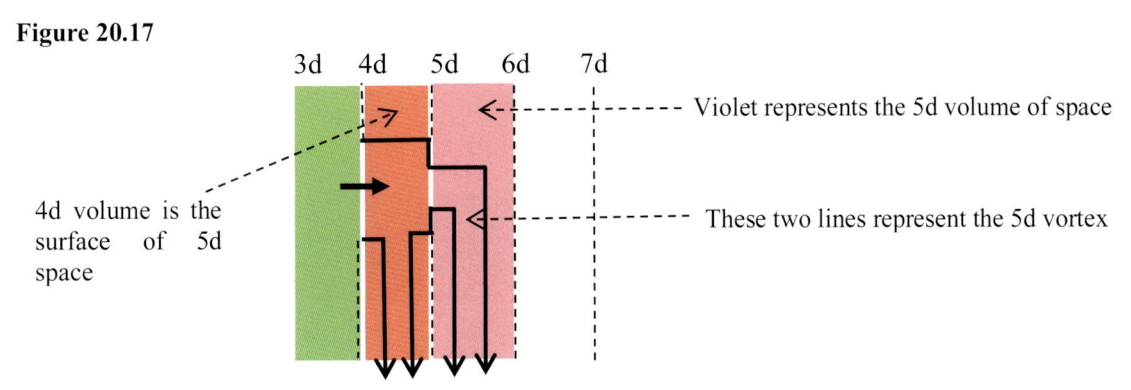

Figure 20.17

The third "layer of matter" starts out with a fifth dimensional surface hole [we call the **tau**] in the sixth dimension volume of space. Notice that the violet 5d volume is actually the surface of the turquoise 6d volume:

Figure 20.18

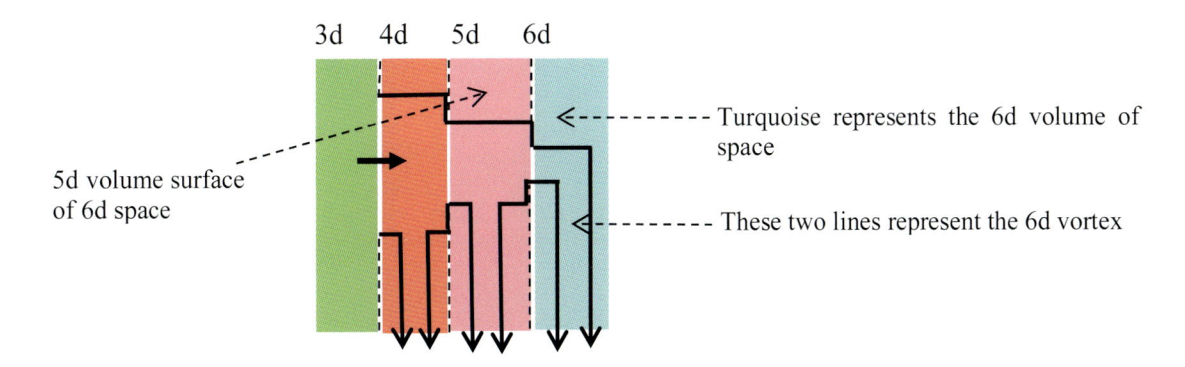

However, unlike the fourth layer of matter seen in the top quark, there is no fourth layer of leptons!

Figure 20.19

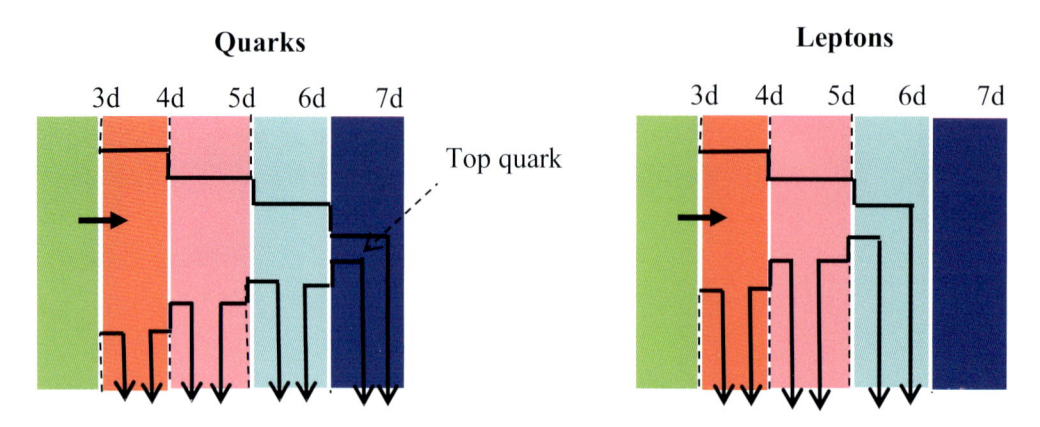

According to the Vortex Theory, because a top quark vortex can exist within the 7d volume of space, a top quark vortex should also be able to exist within a lepton. This fourth lepton would be much more massive than the tau. Because of this characteristic it is appropriately designated **The Wow**! Note too: the Wow would also have a **Wow Neutrino**.

This fourth "layer of matter" would be constructed in the following way: it starts out with a sixth dimensional surface hole in the seventh dimension volume of space. Notice that the turquoise 6d volume is actually the surface of the blue 7d volume.

[Note, update: at the neutrino "telescope" in Antarctica, a fourth neutrino was discovered; revealing that a fourth lepton has to exist!!!]

Figure 20.20

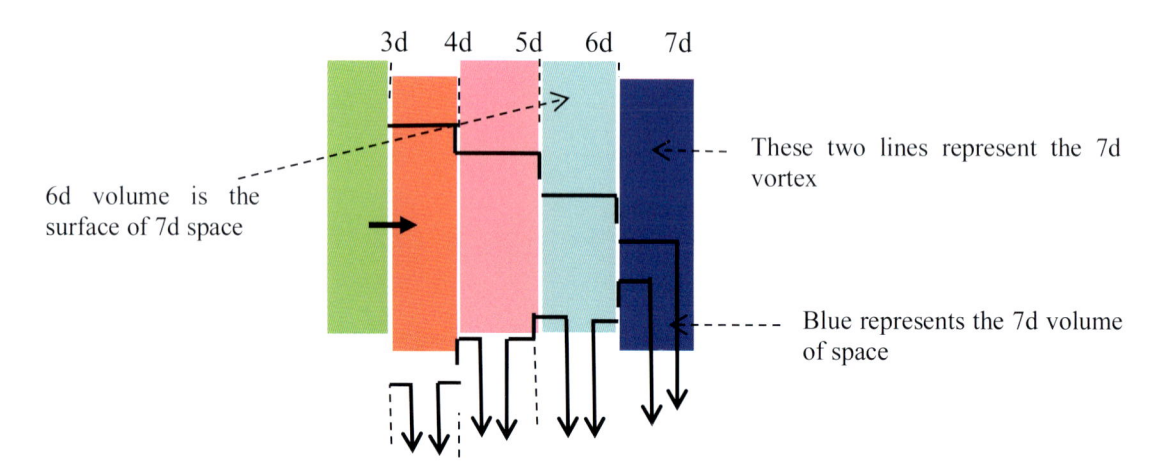

Because of the extremely short lifetime of the top quark [approximately 2.9×10^{-13} s], the decay of the **wow** must also be extremely short. However, short or not, the decay of the wow must follow the decay modes of the tau and the muon. The **wow** must decay into the next lower lepton, the tau and two neutrinos: a wow neutrino, and a tau neutrino.

Because the construction of the neutrino is important to the decay of the wow, the decay mode of the wow will be explained after the section on neutrinos.

Chapter 21
Neutrino Oscillations

Perhaps one of the most mystifying phenomenon in nature are the oscillations of neutrinos and anti-neutrinos. Fantastic as it may seem, these oscillations create an **oscillating mass**! How it happens is most unique. It begins with the distortion of the fourth dimensional plane separating side 1 from side 2.

Three characteristics of the neutrino are responsible for creating the oscillating neutrino mass. The first deals with its compression from the opposite side of space from which it was formed. In the drawings below, notice how the neutrino is bent outward into side 1, while the anti-neutrino is bent inward into side 2. Each of these bends creates stress on the side it is bent into. The distortion of side 1 by the neutrino causes the bent space of side 1 surrounding the neutrino to want to straighten back out; hence it tries to push the neutrino back into side 2; and side 2 will try to push the anti-neutrino back into side 1: See Figure 21.2.

Figure 21.1

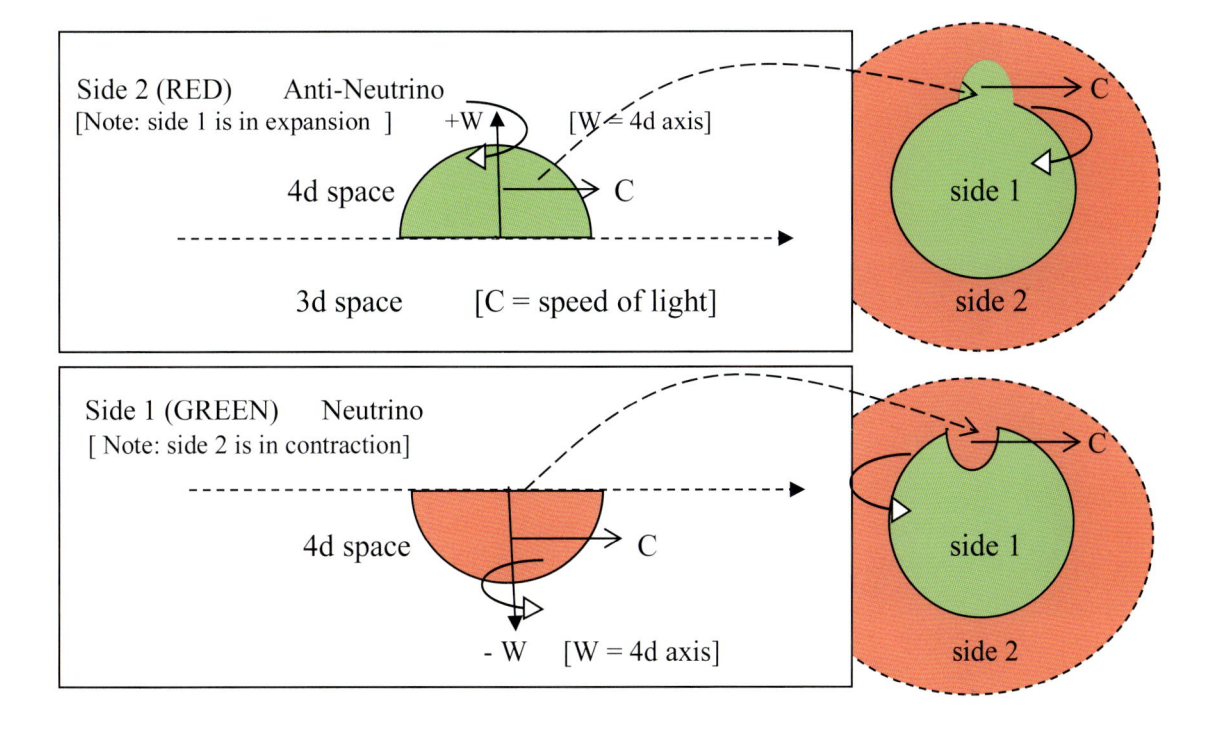

Figure 21.2

Side view: side 2 compresses the anti-neutrino; side 1 compresses the neutrino

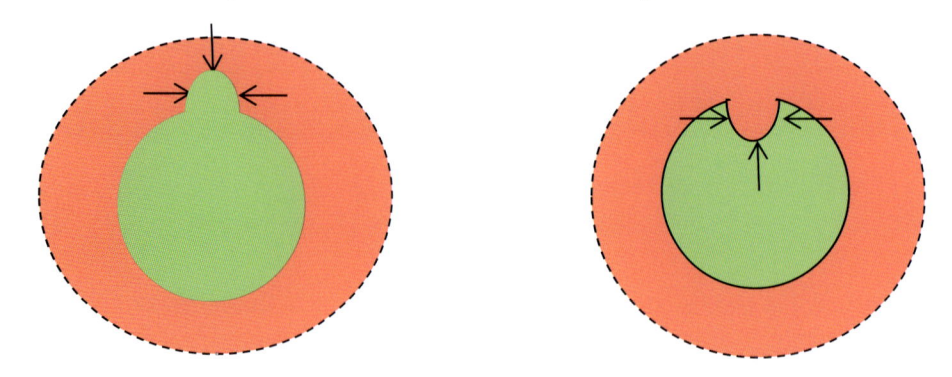

Figure 21.3

Top View:

Figure 21.4

As the pressure increases, the size of the neutrino and anti-neutrino become smaller. However, as this happens, a region of denser space opens up immediately under the anti-neutrino and above the neutrino; while a region of less dense space opens up immediately above the anti-neutrino and below the neutrino:

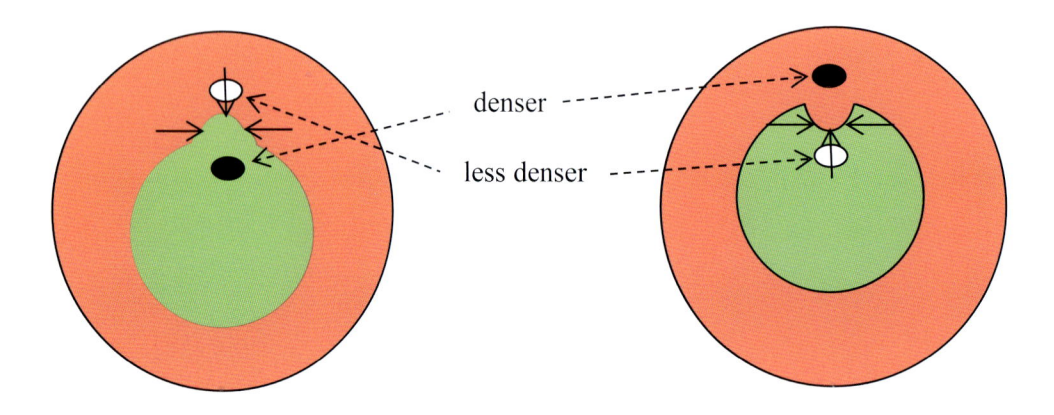

denser

less denser

Figure 21.5

As the neutrino and anti-neutrino continue to become smaller, the denser and less denser volumes of space become larger:

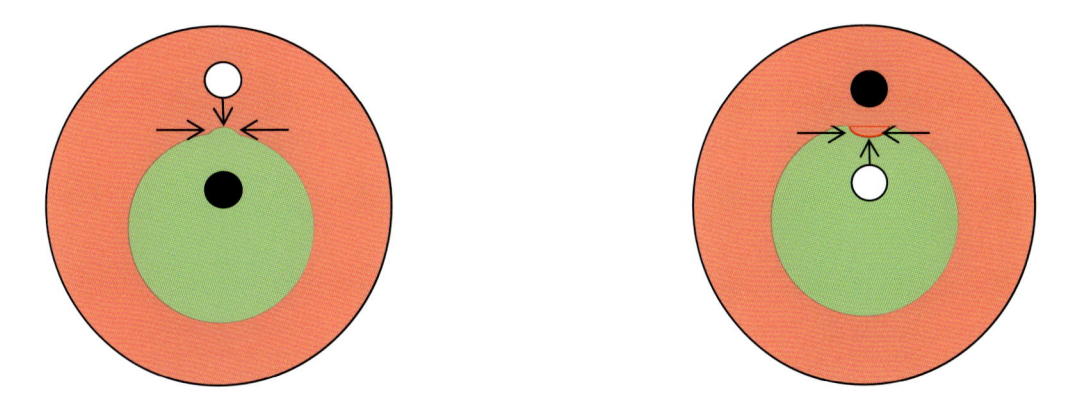

Figure 21.6

Then they disappear, however the denser and less denser <u>volumes of space moving at the speed of</u> the neutrino and anti-neutrino [i.e., the speed of light "C"] continue to exist:

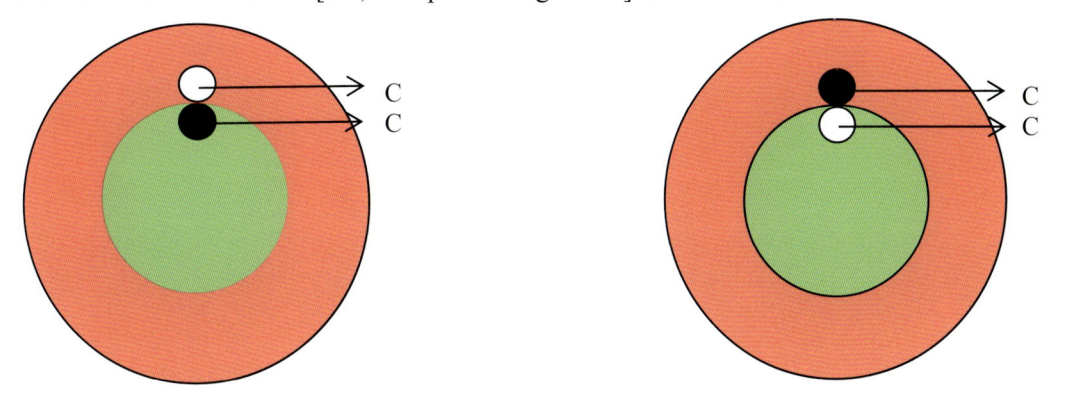

Figure 21.7

  Because the denser regions of space want to expand to recreate homogenous regions of space containing the same density, they take the line of least resistance and begin to move back towards the less denser regions of space above or below them: as this happens, something most fascinating occurs…

Because all four regions are moving at the speed of light [C], they begin to act like photons. And like the photon, they can only contract and expand perpendicular to the velocity of travel. Hence the compressed regions can only expand at 90 degree angles to the velocity of travel. Consequently, the denser region of space finds that it is easier to expand back into the less dense region of space on one side than the denser region of space on the opposite side. Hence, it expands back into the region from which it came…

Figure 21.8

As the denser regions move towards the less denser regions, the neutrino and anti-neutrino begin to reform:

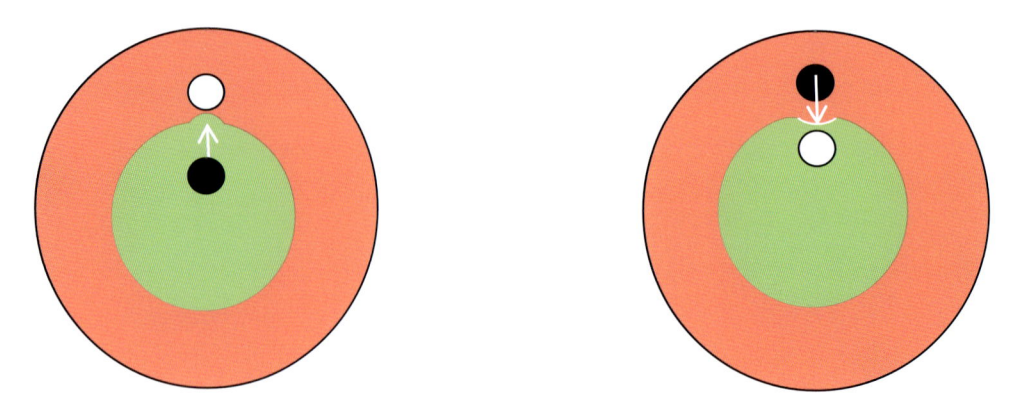

Figure 21.9

The continual movement begins to again recreate the neutrino and anti-neutrino.

denser

less denser

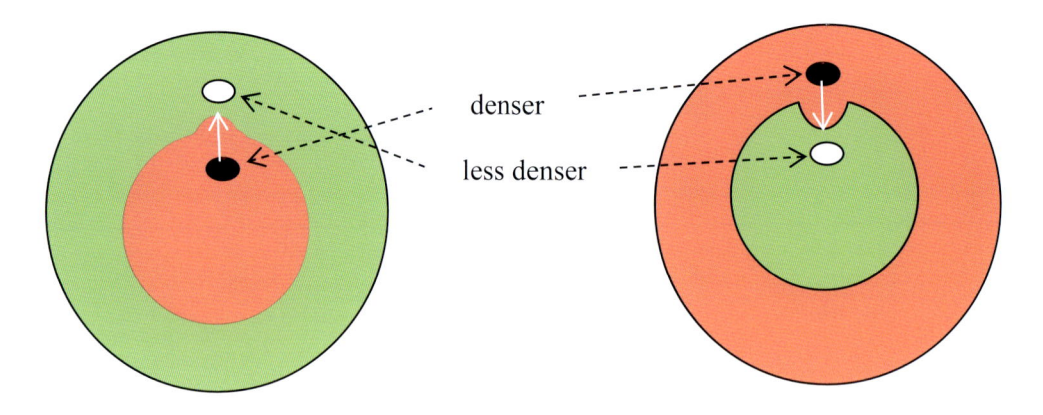

Figure 21.10

The process continues until the neutrino and anti-neutrino are recreated:

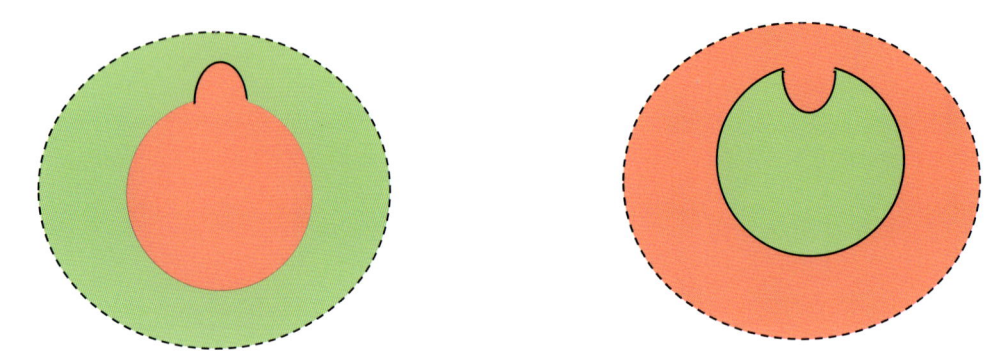

However, as soon as it finishes, it immediately begins all over again.

73

Because the "mass" of either the neutrino or anti-neutrino is at its maximum when the bend of either "particle" is at its maximum creating its maximum mass; and at its minimum when the bend is completely gone, the *mass* of either particle fluctuates between maximum mass and zero mass: a simply remarkable phenomenon unlike anything ever seen in nature!

Figure 21.11

Just like Newton's gravitational constant had to be determined by experimentation, the wavelength, frequency, and amplitude of the electron anti-neutrino and the muon and tau's anti-neutrinos, all have to be determined by experimentation.

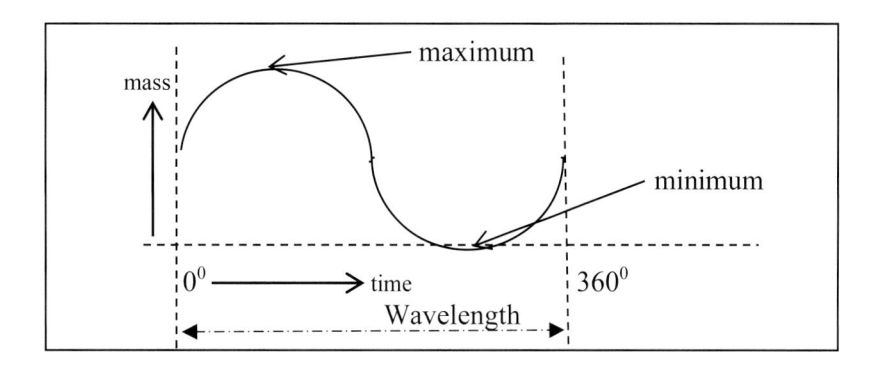

In the past, it was one of the unbreakable laws of particle physics that only an electron neutrino hitting a neutron can create a proton and an electron; that only a muon neutrino hitting a neutron can create a proton and a muon; and that only a tau neutrino hitting a neutron can create a proton and a tau. However, recent experimentation suggests that this is not so. That in fact it appears that an electron neutrino hitting a neutron can create a proton and a muon.

The answer to this strange set of circumstances has been attributed to neutrino oscillations. However, current science does not know how a neutrino oscillates. But now, using the principles of the Vortex Theory, the way a neutrino oscillates is revealed. It must also be understood that if an electron neutrino is expanding outward into 4d space, there will also be a brief but slight expansion into 5d space. It is this expansion into 5d space that makes the electron neutrino temporarily act like a muon neutrino. If a collision happens while this is occurring, a proton muon combination can be formed.

Because the neutrino and the anti-neutrino are formed upon two sides of space, with one side in expansion, and the other side in contraction, both have different elasticizes. These two different elasticizes create different pressures upon each "particle." These different pressures create different expansion and contraction rates in the space in which they exist; making the neutrino and anti-neutrino possess different expansion and contraction rates; different wavelengths and frequencies. The wavelength shown below in Figure 21.12, is shorter than the one in Figure 21.11. If Figure 21.11 was the electron neutrino, then Figure 21.12 would be the muon neutrino. And if Figure 21.11 was the muon neutrino then Figure 21.12 would be the tau neutrino. The greater the mass, the higher the frequency and the amplitude.

Figure 21.12

Again, the wavelength, frequency, and amplitude of the electron neutrino, and the muon and tau's neutrino, all have to be determined by experimentation.

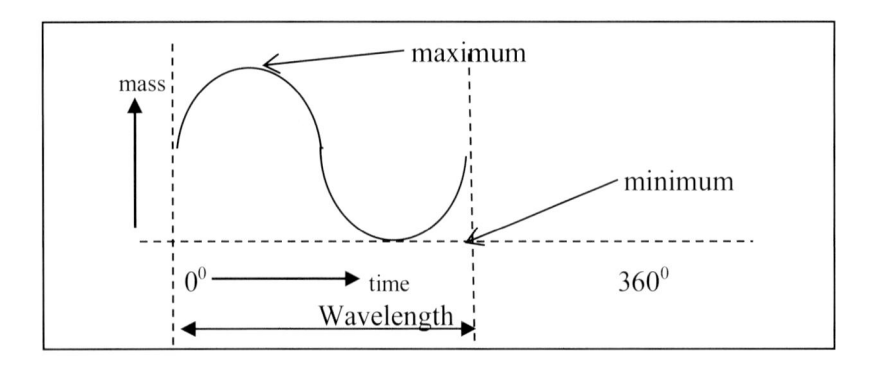

It must also be understood that the above expansion of electron neutrinos and anti-neutrinos takes place between 3d and 4d space, the muon neutrino and anti-neutrino oscillations will take place between 4d and 5d space; while the tau neutrino and anti-neutrino oscillations will take place between 5d and 6d space.

PART II
GREAT PARTICLE MYSTERIES OF THE UNIVERSE FINALLY EXPLAINED

Chapter 22
The Stability of the Proton; The Instability of Mesons

The stability of the proton begins with the realization that like electrons, quarks are also surrounded by regions of dense and less dense regions of space. Just as protons and electrons possess spherical regions of 3d space bent into and out of them, quarks possess spherical regions of 4d space bent into and out of them: [Note, for the sake of simplicity, the terms 1/3 and 2/3 charges are used throughout.]

Figure 22.1

The up quark: because the up quark's charge is +2/3, 4d space is flowing into it. Because space is flowing into it, it is surrounded by a less dense region of 4d space. Since the up quark is of side 1 construction, it is pictured green. Because the anti-up quark's charge is -2/3, 4d space is flowing out of it, it is surrounded by a region of denser space. It is also green because of its side 1 construction:

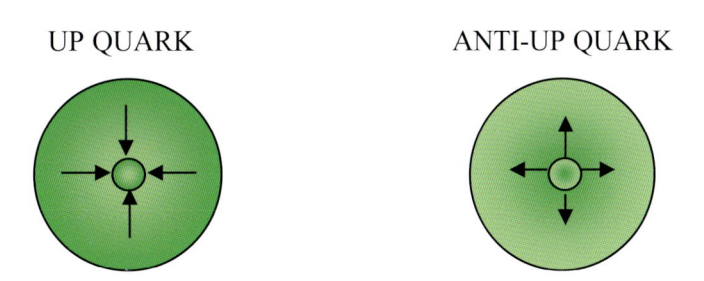

UP QUARK   ANTI-UP QUARK

Because the charm and top quarks are contained within up quarks, their dense and less dense space will also be similar to the up quark; also, because the anti-charm and anti-top quarks are also constructed out of anti-up quarks, they will be similar to the anti-up quark.

Figure 22.2

The down quark: because the down quark's charge is -1/3, 4d space is flowing out of it. Because space is flowing out of it, it is surrounded by a denser region of space. Since the down quark is of side 2 construction, it is pictured red. Because the anti-down quark's charge is +1/3, 4d space is flowing into it. It is also side 2 construction it is also pictured as red:

ANTI- DOWN QUARK   DOWN QUARK

Because the strange and bottom quarks are contained within down quarks, their dense and less dense regions of space will also be similar to the down quark; also, because the anti-strange and anti-bottom quarks are also constructed within anti-down quarks, they will be similar to the anti-down quark.

THE STABILITY OF THE PROTON

The stability of the proton is no accident of nature. This stability is a function of the dense and less dense regions of space created by the proton's two up quarks and one down quark.

Because the two up quarks with their positive charges possess less dense volumes of space bent into them, the down quark possesses a volume of denser space bent out of it, when all three are together within the proton they create a symmetrical "spatial" relationship.

This symmetrical spatial relationship is created with the down quark in the middle and the two up quarks rotating around it. When all three are in this configuration, the dense region of space created by the down quarks is equally balanced on either side by the less dense regions of the up quarks. Having the less dense regions on either side of the down quark, causes its 4d surface to distort outward towards both up quarks simultaneously changing it into a football shape. At the same time, although the surfaces of the up quarks want to distort into pear shapes pointing them and pulling them towards the down quark in the middle of the proton, they cannot.

They cannot because the bent outward space surrounding the down quark is twice as dense as the space surrounding the up quarks. Because this space is twice as dense, the up quarks move outward away from the down quark to a position where the space flowing out of the down quark and flowing into the up quark are the same density. These densities and their corresponding shapes cause the down quark to remain positioned in the center of the proton with the quarks directly opposite to each other: creating a stable configuration. This stable configuration keeps the similar charges of the up quarks from accelerating away from each other - breaking up the proton.

THE INSTABILITY OF MESONS

The reason why mesons are unstable particles comes from the observation that to possess a positive charge, they have to have a quark with a +2/3 charge and a quark with a +1/3 charge. However, because these quarks are both surrounded by regions of bent inward space, they repulse each other. Because there is no third quark to intervene with their repulsion, the more massive quark falls prey to the elasticity of space that wants to contract to alleviate this stressful bend. Hence it contracts causing the more massive quark to decay.

The exact same scenario is true for the negatively charged mesons with their -2/3, and -1/3 quarks.

Chapter 23
The Explanation of the "Conservation" of Strangeness

> Strangeness is a term that developed in the 1950's in an attempt by the physicists of that era to identify reactions that seemed most likely to occur but were not observed to occur. The physicists of that era knew that some sort of new law of nature was involved but they did not know what it was. All they could do then was to label this curious observation "strangeness" and state that it had to be conserved. But what was even more disconcerting to these early particle physicists was the additional discovery that during the weak decay of particles, *strangeness is not always conserved*!

The explanation of the conservation of strangeness is easily explained by the Vortex Theory when it is realized that strange "particles" are really just the two ends of a higher dimensional vortex of flowing space. These two ends exist in two different "particles" of matter such as a negative kaon [K⁻] and a positive kaon [K⁺]. The negative is made up of one strange quark and a negative up quark ($\overline{SU}$); while the positive kaon is constructed out of one negative strange quark and one up quark ($\overline{S}U$).

According to the conservation of strangeness, when numbers of – 1 and +1 are assigned to the strange quarks, it is easy to see that when the two are mutually created in collisions, they will always add up to "0". Also, since they are always created in pairs, when particles collide and create new particles such as those in a linear accelerator, there will never be just one strange quark created. Consequently, the scientists in the 1950s did not understand the mutual creation of particles due to the existence of invisible vortices.

There is also hierarchy of mass in more massive quarks: the strange, the charm, the bottom, and the top. These particles decay slower starting with the strange and faster ending with the top. These faster decays are a result of these holes in space pushing outward on the surrounding. Making the surrounding space denser, making them decay faster and faster.

These densities in space can be visually presented via the following schematics:

Figure 23.1

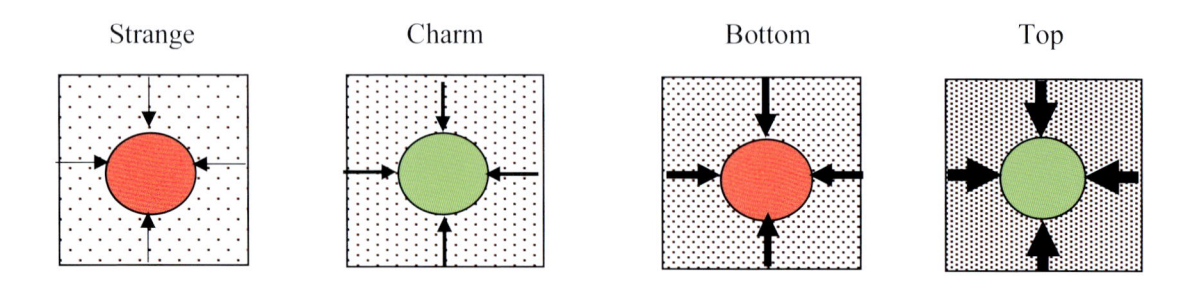

Strange   Charm   Bottom   Top

Notice how the density of the dots representing the density of the space surrounding the quarks increases from quark to quark. This creates increased pressure on the quarks as represented by the thickness of the arrows. Because the strange quark is at the bottom of the list it has a longer lifetime than its more massive associates.

Chapter 24
Gauge Bosons Are *Not* Force Carriers Between Particles!

> According to quantum mechanics, gauge bosons (such as photons) are the force carriers between particles. The photon is the carrier of the electromagnetic force; the gluon is the carrier of the strong force; the W and z are the carriers of the weak force; and the graviton is the carrier of the force of gravity. But this is a mistake. Particles do not carry force between particles. **The particles called Gauge Bosons are not force carriers between forces!**

One of the greatest mistakes in the world of physics is the belief that certain "particles are carriers of force"! Nothing could be further from the truth!

This serious misconception is a result of what can be called: "PARTICLE LOGIC"! This form of logic was unconsciously developed out of Mr. Albert Einstein's outrageous postulate that space is made out of nothing. For if space is made out of nothing, then everything that exists in the universe has to be made out of particles.

As a result of this twisted logic, scientists have spent their time trying to find "particles" (called bosons) that "transfer" force. This has resulted not only in outrageous assumptions (such as the graviton) but a clear misconception in the case of the Higgs Boson!

For example: the graviton…

THE GRAVITON

It is believed that the graviton is responsible for mediating the force of gravity. However, no graviton has ever been seen! This is incredible because the universe should literally be flooded with gravitons. Because every bit of matter that exists is attracted to every bit of matter everywhere in the universe, the universe would have to literally be full of gravitons. In fact, so many would exist that it would almost be impossible to move!

Unfortunately, with the discovery of the Vortex Theory, comes the realization that the forces of nature are created out of bent and flowing space - while the strong force is created by the exchange of up and down quarks between protons and neutrons. So, what are these supposed bosons?

THE MISUNDERSTANDING ABOUT THE PURPOSE AND EXISTENCE OF THE W AND Z PARTICLES

The present day belief that the W and Z particles transfer the weak force is totally and completely wrong. Nothing could be further from the truth.

In comparison to the strength of the strong force, the weak force is indeed "weak". Therefore, it can be said that a "weak" force is responsible for the decay of particles such as mesons, leptons, and baryons. But the W and Z particles are not responsible. It is the different elasticity's of the different dimensions of space that is responsible for the decay of individual quarks.

The W particle is merely a 4d torus containing quarks that is breaking up. Its positive or negative charge is a function of what end is unfolding first [breaking].

The Z particle is a 4d torus caught in a loop.

THE PHOTON

The photon, thought to transfer the electro-magnetic force, doesn't! The photon is a condensed packet of space thrown out of a vortex of flowing space between electrons and protons or discharged from flowing electromagnetic fields by passing matter.

THE HIGGS BOSON

The Higgs Boson doesn't exist. Therefore, it does not and cannot explain mass. Its discovery at CERN was a mistake at best or simply a very convenient misinterpretation of data. Mass is explained when it is realized that space is made of something, matter exists as holes in space, and that their "mass" is a function of the ability of the space surrounding them to create distortions in their surface: allowing them to move as they try to straighten out.

THE GLUON

The Gluon is not a force carrier. The gluon is a higher dimensional vortex of flowing space moving back and forth between quarks.

CHAPTER 23, PART II

This addition to chapter 23 was presented here because it deals with similar subjects…

HOW A GAMMA RAY CREATES AN ELECTRON POSITRON PAIR

The gamma ray is a high density photon [big photon containing a large volume of dense 3d space]. When gamma rays collide, their rapid expansion and contractions are so swift they create a rip or tear in 3d space.

As the tear separates, space flows into one side and out of the other creating the electron positron pair. Because the photons were moving at the speed of light, the tear was created at light speed and space flows into and out of the holes at light speed.

BETA PARTICLE

The release of an electron from within the nucleus of an atom is called Beta Decay. This release is accompanied by the change of a neutron within the nucleus into a proton.

The explanation is simple. The 4d torus creating the neutron breaks; one end is the proton, the other is the electron. The proton stays within the nucleus while the electron is thrown free.

THE EXPLANATION OF THE ALPHA PARTICLE

One of the great observation achievements of early particle scientists was the discovery that only three types of particles are expelled from the nucleus of an atom: alpha particles, beta particles, and gamma rays.

Now the gamma rays don't present a problem because they are nothing more than energetic photons; nor do the beta particles because they are merely electrons. However, the alpha particles –which are nothing more than helium nuclei – do present a very special problem: why only alpha particles? Why are only alpha particles thrown out of the nucleus? Alpha particles consist of two protons and two neutrons. So, what is so special about this arrangement? Why don't we see particles made up of one proton and one neutron, or three protons and three neutrons, or four protons and four neutrons, or more?

The answer is found in the true nature of the strong force. The strong force is not an attraction between particles, but rather the continual transformation of a proton into a neutron and a neutron back into a proton. This process is easy to observe between particles that are paired with each other, but it is also easy to observe in two pairs of particles.

When a proton neutron pair is close to another proton neutron pair, the continual transformation keeps all four particles "stuck" together:

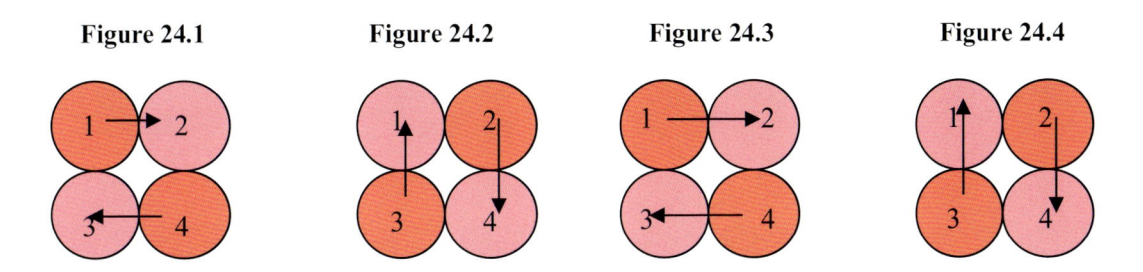

Figure 24.1   **Figure 24.2**   **Figure 24.3**   **Figure 24.4**

[note: red = proton: lavender = neutron]

In the above sequence note how 1 becomes 2, 2 becomes 4, 4 becomes 3, and 3 becomes 1 completing the circuit. Note also, that if 1 became 2 as 4 became 3, and then 2 became 1 again as 3 became 4 again each pair would be acting as a separate set instead of one set. In this circumstance, set 1&2 and 3&4 would be separate from each other and would not be held together by the strong force. Instead, the two sets would be pressed together inside of the nucleus due to the bent inward regions of space that surround each particle.

TWO NUCLEAR "FORCES"…NUCLEAR GRAVITY

Two "forces" appear to be at work inside the nucleus. The first is the bent inward regions of space surrounding each nucleon. This bent inward region surrounding each nucleon is the force that initially attracts the neutron to the proton. The strong force between protons and neutrons [see Chapter 33] is what holds them together. This unrecognized force can be called "NUCLEAR GRAVITY"!

THE INSTABILITY OF ISOTOPES

The excess neutrons that create isotopes of a particular atom are unstable because the switching process that takes place back and forth between protons and neutrons must now encompass more neutrons. The roundtrip that usually takes place between two protons and two neutrons must now take place between at least one or more neutrons. This creates an unstable situation because a neutron is allowed to remain a neutron for a longer period of time. This allows the space within the neutron to circulate for a longer period of time allowing the torus that creates the neutron to decay faster.

Chapter 25
The Explanation of the Pauli Exclusion Principle

> Using the principles of the Vortex Theory, the construction of the alpha particle, and the theory that the nucleus is constructed out of alpha particles, the explanation of the Pauli Exclusion Principle is explained. If protons and electrons are connected to each other via fourth dimensional vortices, they spin in opposite directions. Since the alpha particle possesses two protons possessing opposite spins, their electrons also possess opposite spins. With a nucleus constructed out of alpha particles, all paired electrons in shells and sub-shells will spin in opposite directions.

HISTORICAL BACKGROUND

In 1925, Wolfgang Pauli after a long and arduous effort to explain the so-called "anomalous Zeeman effect" finally discovered what has come to be known today as the Pauli Exclusion Principle. This principle states that no two electrons in the same atom can have identical values for all four of their quantum numbers. For example:

If two electrons occupy the 1s orbital, the first three quantum numbers of both electrons are identical: $n = 1$, $\ell = 0$, and $m_\ell = 0$. Since all three of these numbers are the same, the Pauli Exclusion Principle states that their fourth quantum number, their spin number, has to be different: one electron must have a spin of $m_s = +1/2$, and the other have a spin of $m_s = -1/2$. Consequently, what the exclusion principle reveals is that two electrons in the same orbital must have opposite spins.

Unfortunately, although this brilliant deduction by Mr. Pauli explained the Zeeman Effect and eventually won him the Nobel Prize in physics, there has always been one major problem with this great discovery – it is an empirical relationship.

An empirical relationship is based upon observation rather than theory; there is no theoretical reason to believe in the relationship, only data reveals it to be so. And so it is with the case of the Pauli Exclusion Principle. For the past 80 years, nobody including Pauli himself knew what was creating it. It has been an observation without an explanation. Only now, with the discovery of the Vortex Theory does the answer finally become apparent.

Because the principles of the Vortex Theory are used to explain the Pauli Exclusion Principle, a brief summary of this new and revolutionary theory is given:

THE VORTEX THEORY

According to the Vortex Theory, the proton and the electron are connected by a vortex of three dimensional space flowing from the proton to the electron in higher dimensional space!

Figure 25.1

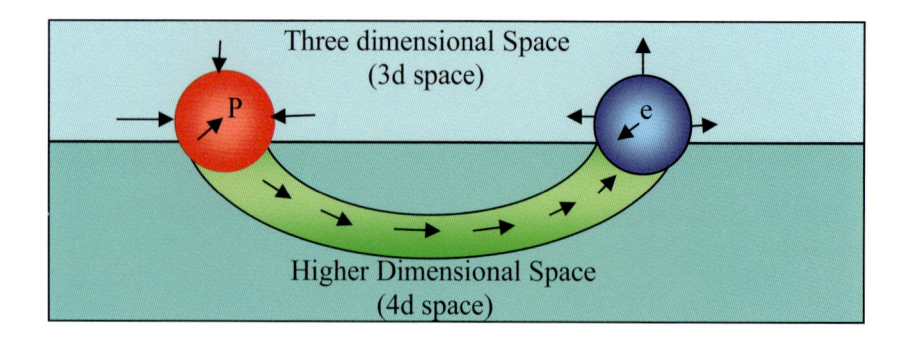

When a hydrogen atom is created, some of the space flowing out of the electron begins to flow into the proton; as these two holes move closer together a critical distance is reached where all of the 3d space flowing out of the electron flows directly into the proton. When this situation occurs, *a second* vortex of whirling space is created. These two vortices create a circulating flow containing a fixed volume of space.

This circulating volume of 3d space continually flows from the proton, into 4d space - through 4d space, and then into the electron. Here, it exits the electron, flowing back through 3d space and into the proton once again, binding the proton to the electron creating a hydrogen atom.

Figure 25.2

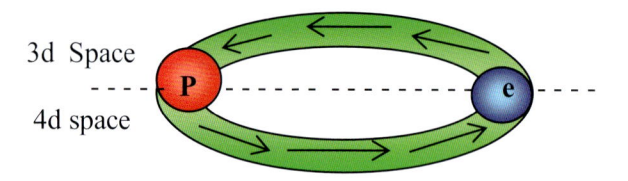

When the circulating flow commences, both of the electrostatic charges are neutralized. The word "neutralized" was used because no flowing space escapes from the system. If surrounding space still flowed into or out of this system, all atoms would possess electrical charges (and every time we touched something we would get shocked). Note: ions are created when a molecule is broken up and a proton in one atom is connected to an electron in another atom via a 4d vortex.

[Also: it is necessary to mention that the neutron is a vortex caught in a loop. The neutron is a proton completely surrounded by an electron: a hole within a hole. As such, it is turned into a 4d torus [a 4d vortex turned inside-out (similar to a smoke ring though impossible to draw)].

A CURIOUS OBSERVATION

The bases of the Vortex Theory now allow us to explain the Pauli Exclusion Principle. An explanation that begins with a most curious observation regarding the nucleus of the atom: why does the nucleus emit alpha particles? The alpha particle is a helium nucleus containing two protons and two neutrons: but why this combination? Why only two protons and two neutrons?

Why aren't particles emitted that contain three protons and three neutrons; or four protons and four neutrons? Also why is the number of protons and neutrons equal? Why aren't particles ejected that contain two protons and one neutron; or two neutrons and one proton?

THE ALPHA PARTICLE

The answer to the above question is found in the unique structure of the helium nuclei. This uniqueness is first found and characterized by its neutral spin. This neutral value reveals that the spins of all four nucleons cancel. And if the vortices exist, because the spin states of the electrons in a helium atom are opposites, the spin states of the protons are opposites; also, to create neutrality, the spin states of the neutrons must cancel. It is this neutrality that appears to be holding the alpha particle together.

According to the Vortex Theory, the intrinsic magnetism of fermions is caused by the rotation of the space around them; [fermions are spinning around a fourth dimensional axis]. Opposite spin states create opposite rotations creating opposite polarities. Because the 2 protons and the 2 neutrons in the alpha particle possess opposite spins they possess opposite polarities. These opposite polarities cause the 2 protons and the 2 neutrons to be attracted to each other much like *Cooper pairs*; binding one proton neutron pair to the other – binding the alpha particle together.

The alpha particles in the nucleus of the atom are bound together by the less dense space surrounding protons and neutrons. According to the Vortex Theory, the accumulation of all the less dense volumes of space surrounding all the protons and neutrons in a planet or star is responsible for creating its gravitational field. On the subatomic scale, these same volumes of less dense space are responsible for creating "nuclear gravity" that bind the alpha particles together in the nucleus, and also the excess neutrons responsible for creating isotopes of a particular atom.

Surprisingly, using the principles of the Vortex Theory, the explanation for Pauli Exclusion Principle can now be illustrated; [although the spin states of each type of particle is opposite, note particularly the spin states of the electrons]:

Figure 25.3

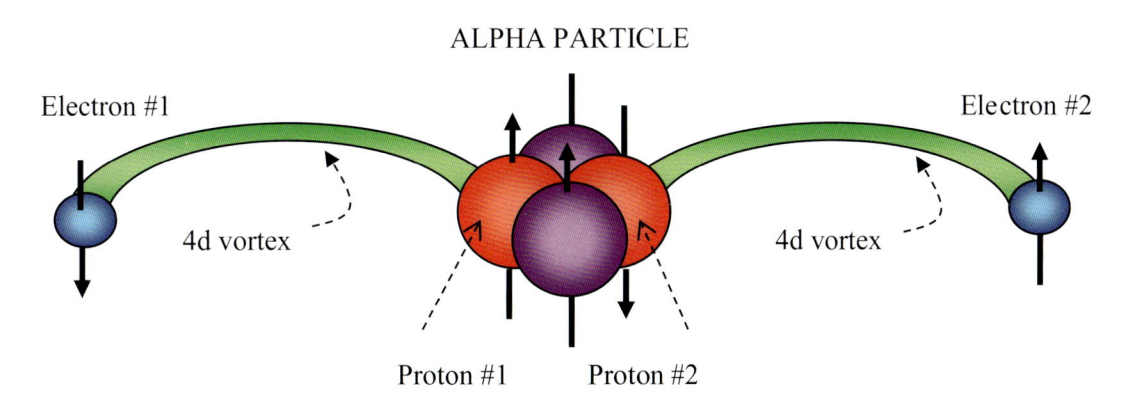

ALPHA PARTICLE

Note how the spin states of the protons in the alpha particle are opposite to each other. Although the spin states of the proton and the electron it is attached to are also opposites, because each is at the other end of the rotating vortex, the spin states of the two electrons are now opposite to each other; consequently, *the spin states of the two electrons are opposite to each other because the spin states of the protons they are attached to are opposite to each other.* This is the simple yet elegant explanation for the Pauli Exclusion Principle.

Figure 25.4

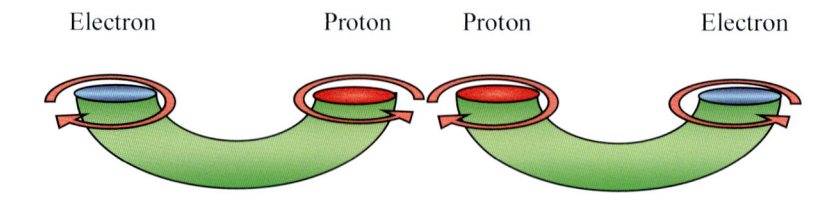

Electron　　　　Proton　　　　Proton　　　　Electron

NOTE: the explanation of how the electron can have its spin reversed during a collision with another particle then flip back to its original spin state creating the 21 centimeter line in astronomy is the subject of the next paper on the Pauli Exclusion Principle.

Chapter 26
Explanation of the CPT Theorem

> The C- charge conjunction, and P- parity (space reversal) of the CPT theorem can be easily explained by the vortex that connects a particle with their anti-particle. The T- time reversal presents a problem most scientists have never before considered. A discussion follows:

According to the model of the universe proposed by the Vortex Theory, the charge a particle possesses is created by the space flowing into or out of a particle. For example, in the proton anti-proton pair, three dimensional space flows into the three dimensional hole of the proton then outwards into fourth dimensional space, through fourth dimensional space, into the anti-proton where it again exits back into three dimensional space. This relationship can be seen in the 3d to 4d relationship drawn below:

Figure 26.1

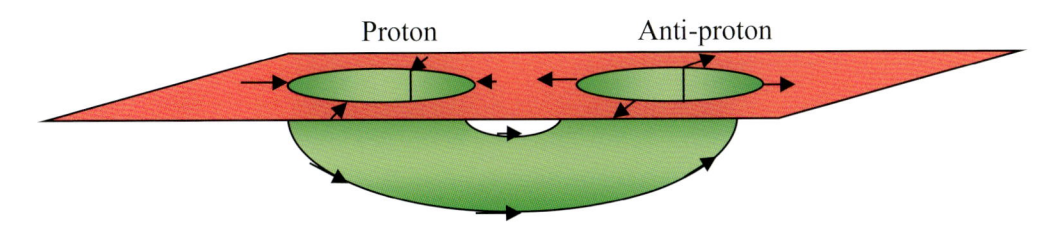

In the above drawing, notice how the three dimensional [3d] space flows into the proton through the fourth dimensional vortex then back through the anti-proton and out onto the 3d plane. Note too, how the 3d space is really the surface of a volume of 4d space contained within the vortex.

CHARGE CONJUNCTION

Applying the above relationship to 3d and 4d space, since the 3d space that flows into one "particle" flows out of the other "particle", it can be seen that the Charge Conjunction is a result of the 3d space flowing between the two particles in the form of a vortex.

PARITY

According to the model of the universe proposed by the Vortex Theory, the isotropic spin of "particles" [holes in space] is a function of the direction of their orientation with regards to 4d space. Because it is impossible to draw 4d space, the 3d to 4d relationship is again drawn:

Figure 26.2

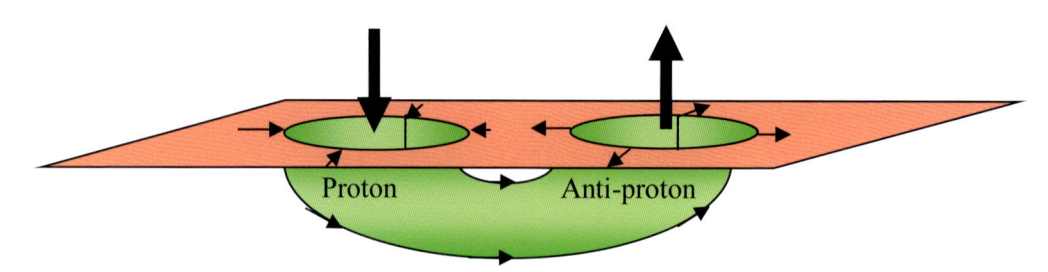

In the above drawing, note how the 3d [three dimensional] space [thick black arrows] is flowing into the proton and out of the anti-proton. Note too how they are in opposite directions to each other – creating the opposite effects.

And again, applying the above relationship to 3d and 4d space, it can be seen that when the space within the vortex is rotating, opposite spin states are created at the opposite ends of the vortex. Because the vortex ends are connected to the same surface, a clockwise rotation in one end produces a counter-clockwise rotation in the other end. For example:

Figure 26.3 [Top view of Figure 26.2]

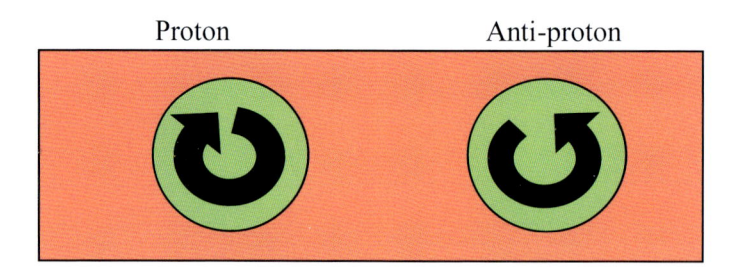

In the above drawing it can now be seen how the rotation of space within the vortex creates the opposite spin states of a particle and its anti-particle. Also, because the vortex model of the universe proposes that magnetism is nothing more than the rotation of space, it can now be seen how opposite magnetic moments of the two "particles" are also created."

However, even though the space reversal – parity invariance – is symmetrical, there is one characteristic of the two particles that is not symmetrical: its gravity!

The Vortex Theory of matter reveals that gravity is created by a less dense region of space surrounding protons and neutrons. This region of less dense space is created by the flow of space into the proton [the neutron was discussed earlier]. However, because space is flowing out of the anti-proton, a denser region of space now surrounds it. And even though this dense region of space mirrors the <u>exact opposite</u> symmetry of the proton, the dense region also creates the exact opposite effect regarding gravity: it creates an anti-gravity effect! In essence the anti-proton is an anti-gravity particle.

This characteristic of the anti-proton [and anti-neutrons that are created out of anti-protons] is the reason why we do not see clusters of anti-matter protons and neutrons. An anti-proton, positron hydrogen atom is an anti-gravity atom. Although its anti-gravity effect is probably too weak to be measured by today's instruments, a cluster of anti-matter hydrogen atoms should create a most interesting effect. If somehow contained within some sort of special container, they would all want to move away from each other. Hence, they will be found at the inside edges of the container instead of in the middle of it.

[Note too: during the early universe, the anti-gravity effects of anti-matter apparently kept it isolated from the other matter of the universe and kept it from annihilating with the atoms. Note, the antigravity effects of anti-hydrogen atoms made of anti-protons and positrons would have kept them isolated from today's hydrogen made of protons and electrons. Note too: most of the anti-matter resides in the vast voids between the Galaxies creating what has been called Dark Matter.]

TIME REVERSAL

One of the great discoveries of the Vortex Theory and proved in the vortex thesis is the revolutionary idea that time is a function of motion, a phenomenon created by motion and cannot exist apart from motion. As such it does not exist as a fundamental principle of the universe. Therefore, it cannot be reversed.

The Vortex Theory reveals that time does not exist as a fundamental principle of the universe and cannot be reversed, because the motion of the matter of the universe cannot be reversed backward to an "earlier time"!

CPT VIOLATION IN PION DECAYS

The non-conservation of charge conjugation in the weak decays of pions is a pivotal clue to the manner in which the entire universe itself was created. This clue begins to gain prominence when it is used to explain the asymmetric parity of neutrinos.

When the asymmetric parity of neutrinos was first proposed by Tsung Dao Lee and Chen Ning-Yang in 1956, it was ill received – especially by Wolfgang Pauli. The idea that all neutrinos appear to be "left-handed" [spinning counterclockwise] while all anti-neutrinos appear to be "right handed" [spinning clockwise] seemed to defy logic.

But then, in 1957, when Chien-Shiung Wu proved it was true, the world of physics was stunned, (and Pauli had to apologize). However, even though it is sixty years later, this discovery still seems to defy all logic. Because why should neutrinos only spin counterclockwise and anti-neutrinos only spin counterclockwise when hadrons, mesons, and other particles such as electrons and positrons can all spin either clockwise or counterclockwise?

The answer, like all other answers, can be explained using the correct vision of the universe. The answer only seems illogical when men try to explain it using the wrong vision of the universe.

The answer begins when it is realized that there are massive differences in how neutrinos and charged "particles" are created. Unlike charged particles, neutrinos and anti-neutrinos do not penetrate into the interior of space. Instead, they are quantized transverse waves created upon the surface of space. Because there are two volumes of space, one in expansion, one in contraction with each having the opposite spin to the other, neutrinos take the spin of the *surface of the particular volume of space they are formed upon*. Although electrons and positrons appear to be surface phenomenon they are not. They are the three dimensional ends of vortices of flowing space moving into and out of the interior of higher dimensional space. This is where they differ from neutrinos and anti-neutrinos. They are affected by the ability of the volume to move, and not by the limited ability of the surface to move.

Figure 26.4 The neutrino and anti-neutrino as seen from Side 1 and Side 2

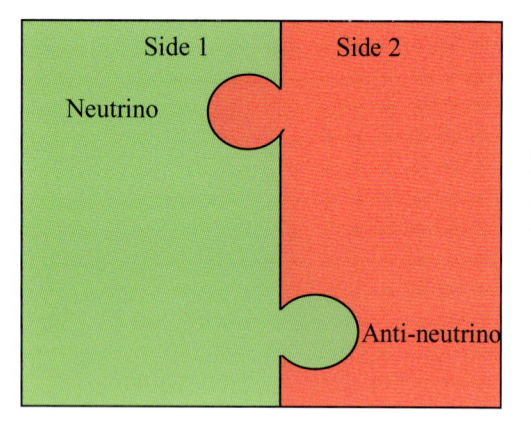

Note: because the neutrino and the anti-neutrino are quantized waves, they are Spherical or "bubble" shaped

[Note: how the neutrino is bent outward, into side 1, while the anti-neutrino is bent outward into side 2.]

Because neutrinos and anti-neutrinos are transverse waves existing upon the surface of the two volumes of higher dimensional space, their motions [such as their asymmetric parity] are determined by the surface characteristics of side 2 and side 1. This is an important observation because it not only allows us to explain how neutrinos and anti-neutrinos are constructed, it also gives us a glimpse into how the universe itself is constructed.

For example, if the universe was created by a region of higher dimensional space turned inside-out creating two volumes of space [one in expansion and one in contraction] then two characteristics emerged that would create the asymmetric parity of neutrinos: #1 the isotropic spin of the universe; #2 the principle of left becoming right.

THE ISOTROPIC SPIN OF THE UNIVERSE

Because of the vast and immense size of the physical universe, the notion that the *space* of the universe exists as one single particle seems hard to believe. However, there is good reason to believe it is so.

If the universe was created by a region of space turning inside out, at the point of creation, this volume would make its appearance as a spinning microscopic sphere expanding outward at the speed of light. If the Vortex Theory's definition of isotropic spin is true, then spin up and spin down are defined as directions into or out of higher dimensional space. Therefore, the bend outward surface of the expanding sphere would possess "spin up" and the surface of the contracting sphere would possess "spin down;" making the spin of each surface the exact opposite to the other.

The penetration of one volume into the other would decrease the volume of the contracting volume, increase the volume of the expanding volume and allow one to expand and the other to retreat at the speed of light.

However, because *momentum is a characteristic of mass*, even though the 3d surface of the universe would be massive in size in relation to the original tiny hole that began the expansion, it would not possess momentum so it could not lose momentum. Hence, it would not lose "angular momentum." Its expanding surface [side 1] would still possess its helicity of – 1, $m_s = (-)$ making an indentation pushed from side 2 into side 1 twist to the left - making all neutrinos left handed; while (side 2) would possess a helicity of +1, $m_s = (+)$ making an indentation pushed from side 1 into side 2 twist to the right - all anti-neutrinos right handed [as will be explained shortly].

Parity of spin: because spin is a characteristic of the surface of a particle, the two sides of space are in fact the surfaces of two different massive particles. Furthermore, since one volume is turned inside out in relation to the other volume, its spin will be opposite. So, if the expanding volume of space was spinning in a clockwise motion; then from its perspective, the other volume of space would appear to be spinning in a counterclockwise motion. So if a portion of the expanding [Side 1] surface [in the form of a spherical "bubble"] is pushed outward into the opposite or contracting volume [Side 2] creating an anti-neutrino, this spherical bubble will be forced to take on the spin of the opposite volume: it will seem to spin from right to left or counter-clockwise from our point of view. Furthermore, if a portion of the contracting volume was pushed outward into the expanding volume creating the neutrino, it will be twisted from right to left and forced to take on the spin of the opposite volume.

WHEN LEFT BECOMES RIGHT

The vision of space turning inside out is another principle possessing profound consequences. If the physical universe was created by a region of space turning inside out, and expanding into the volume it originally was a part of, then two different volumes and two different surfaces of space would be created. Furthermore, if one volume is turning inside out, everything in it is reversed as previously explained.

IN CONCLUSION

Although the CPT Theorem does not realize that negatively charged particles create anti-gravity effects, the theory is still valid in regards to both charge and spin. This is true because even though gravity and anti-gravity are in effect the mirror images of each other, when the direction of space is reversed in a charged particle, and space flows out of an anti-proton instead of into it, a bent outward region of space is created around it instead of the bent inward region. The anti-proton would now produce anti-gravity effects.

For these two particles as well as all hadrons and charged leptons, when their spatial vectors are reversed, not only is their symmetry reversed, but their gravitational effects are also reversed: confirming this part of the CPT Theorem. However, this theory is still stuck with the problem of time.

Because time does indeed appear to be a function of motion - a phenomenon created by motion - then in *actuality,* like entropy, the arrow of time cannot be reversed: it has no opposite vector. And the theory is invalid. But then again, this failure only exists when time is used in actuality; if the time component of the CPT theorem is only used *hypothetically* [as in an imaginary reversal] then its use again becomes perfectly acceptable. And the theory is valid.

The only problem with this theory concerns the neutrino. However, once it is understood that the neutrino is a specially constructed "particle," the problem is resolved.

Chapter 27
Motion of Photons and Particles Through Electric and Magnetic Fields

> Question: if magnetic fields and electrostatic fields are really flowing space, then when a photon encounters a magnetic or electric field, why doesn't the flow change the trajectory of the photon? Example, when a man tries to swim directly across a river, when he reaches the other side, he finds that he is downstream from the place where he dove in. So, it is only logical to assume that when a beam of light from let's say a flashlight, is directed at the edge of a magnet or an electrostatic charge, the photons should be deflected. So why aren't they?
>
> The answer is found in the density of space.

When a flashlight is shined perpendicular to an electrostatic charge or a magnetic field, it reacts to the flow of space exactly as if it were a denser region of space. Because a denser region of space is able to bend and flex faster than a less dense region, the photon's velocity increases in the direction of the flow. Hence, it moves at a faster speed at an angle to the flow.

This faster speed and different angle causes it to move through the flowing space as if it were still on a straight trajectory; allowing it to pass through the field as if the field was not there. The same is true for a magnetic or static charge flowing directly at the beam of light. In this situation, the space directly in front of the beam of light appears to have increased to twice its density, allowing the photons to travel at twice the speed of light! [We will address this subject in the next chapter.]

Also, when traveling in the same direction as a magnetic field or electrostatic field, from the point of view of the photon; the density of the space in front of it is almost nil: creating the reverse effect.

Although at first appearance, the above explanations appear to be "ad hock" this is not so. The above explanations were derived from the relativistic effect created by the speed of light in an intense gravitational field.

In an intense gravitational field, all motions slow down. This effect is created by the slower motions of all of the atoms in the field that in turn slow down the motions of all objects in the field; slowing down clocks: making it appear as if time has slowed down. However, and this is most important, if the speed of light has not also slowed down within this intense gravitational field, the measurement by the slower moving clocks would make it appear to speed up. Since this does not happen, and since the intense gravitational field is actually a region of less dense space, the speed of light is slowing down in a region of less dense space.

COLLISION OF TWO PHOTONS IN SPACE

Another problem encountered by the motion of a photon is the encounter with another photon. When the encounter occurs, because they are both regions of dense space, why doesn't one ricochet off the other? Or why don't they somehow explode into each other?

The answer to the first question is again explained by the increased speed of a photon when encountering a denser region of space. Because each photon perceives the other as a region of increasing density, each one speeds up as their denser regions of space approach each other and pass through each other. Likewise, as they separate, the attempt to bend back into each other slows them down and returns them to their normal speed.

MOTIONS OF PARTICLES THROUGH ELECTRIC AND MAGNETIC FIELDS

Again, if electric and magnetic fields are flowing currents of space, then why aren't atoms deflected by them? Like the swimmer in the river, why aren't they caught up in the flow?

The answer is most unique. It begins with the fact that the protons and electrons that atoms are made out of are not "particles" they are holes in three dimensional space.

Because they are holes, they encounter an entirely different set of circumstances when placed within a region of flowing space. Since they are not particles, space does not "push against them" and carry them along with the flow. Instead, space has to flow around them.

When space flows around them, the flow has to split apart. As it initially splits apart, the side of the hole in the direction of the flow is distorted outward into a pear shape. This pear shape tries to return to the shape of a sphere, it continually moves, accelerates, moves and accelerates in the direction of the flow. Since the pear shape is proportional to the volume of the flow [the intensity of the field], a greater volume creates a greater bend, creating a greater acceleration.

This relationship between the distortion and the field creates an inverse relationship between the strength of the field and opposition to its motion. Hence, the atom does not move with the field but in fact is actually moving against it, creating a stalemate.

Chapter 28
Why the Electron Can Tunnel but Ordinary Matter Cannot

> The reason why the electron can tunnel through the wall, but a baseball cannot, is explained by the vortices. The free electron is at one end of a vortex of flowing space that is not connected to another particle moving with it. Consequently, when the electron reaches the wall, it can drop into 4d space and continue on its way; but an electron that is connected to a proton in an atom cannot.

It cannot because the space flowing out of the electron is trapped within the three dimensional vortex flowing back to the proton in 3d space. Consequently, this three dimensional volume of flowing space cannot dip down into 4d space; keeping the electron trapped upon the 3d surface. Hence, although a single electron can tunnel, an electron in an atom cannot.

Chapter 29
Eliminating Popular Misconceptions

> The truth about certain misconceptions and mythical particles: GLUONS, Gravitons, Higgs Boson Particle, Sea of Quarks, Sea of GLUONS, Magnetic Monopoles, Color Changes…

Some of this information was previously presented. It is re-presented here because some of this false information possesses other conflicts not explained before.

STRONG FORCE AND GLUONS

Gluons are not carriers of the strong force.

If gluons were responsible for the strong force, protons would not be a necessary ingredient needed to create the nuclei of atoms. Neutrons would "clump" together. We would see massive nuclei made up of large numbers of neutrons. Since there would be few protons to attract electrons, these clumps would be the equivalent of microscopic rocks!

Because such particles are not seen, something else must be responsible for the creation of the strong force. Hence, the strong force created by a virtual particle being passed back and forth between protons and neutrons is explained in this book.

GRAVITONS

Gravitons do not exist. The force of gravity is not carried by the graviton. Instead, the force of gravity is created by a distortion upon the surface of a three dimensional hole in space. [All is explained in Book 2.]

HIGGS BOSON PARTICLE

The Higgs Boson particle does not exist. Mass is not transferred from one particle to another by an intermediary particle. Instead, mass is a phenomenon created by the ability or inability of a surface to distort. The harder it is to distort the surface, the more mass a "Particle" appears to have. [Also explained in Book 2.]

SEA OF QUARKS? SEA OF GLUONS?

It is currently believed that quarks within hadrons are "bathed" in a "sea of gluons" and additional quark anti-quark pairs. But this is a mistake. The 4d space within the hadron is made up of a substance that gluons and quarks are *created* out of.

In an effort to explain how quarks change "flavor" or how different types of quarks are created when quarks decay or when quarks are created during collisions, the idea of a sea of gluons was proposed by Paul Dirac a great mathematician but was mistaken about the construction of matter. Consequently, this idea that space is made up of a "sea of gluons" is a mistake. With the discovery of the Vortex Theory and the realization that matter and anti-matter are really just the ends of miniature vortices, the idea of the "sea of quarks" is revealed to be an error.

MAGNETIC MONOPOLES

Another mistake; The Vortex Theory reveals that the magnetic force is being generated by the rotation of 3d space around holes in space, and not being generated by "particles". Hence, magnetic monopoles do not exist.

COLOR CHARGES

The explanation of the Pauli Exclusion Principle [in this book] reveals the true explanation for the electron's quantum numbers; that it is the position of the proton within the nucleus that creates a different set of numbers for each electron; and that the electrons quantum numbers are an effect and not a cause.

Because of the mistaken 20th Century reasoning regarding the electron's quantum numbers, when the good intentioned scientists of the past tried to apply the Pauli Exclusion Principle to quarks, they naturally believed quarks would also possess their own set of quantum numbers that had to be different from each other. Therefore, they developed the theory that there had to be three different types of charges or "colors" for quarks. Ironically, they were correctly mistaken! Although there are in fact three types of charges; the correct explanation for the Pauli Exclusion Principle now eliminates the need for three "color" charges.

Chapter 30
Major Problems With Today's Popular Theories

> THE STRING THEORY, TIME, QUANTUM CHROMODYNAMICS, CHARGE QUANTIZATION, THE STANDARD MODEL AND GRAVITY; MASS; all have major problems!

THE STRING THEORY – AND TIME DILATION

Although the string theory is beautifully constructed out of elegantly vibrating one dimensional strings, it is nevertheless based upon the existence of Einstein's fourth dimension of space-time. It needs this fourth dimension of time to be able to explain length shrinkage and time dilation. But if time does not exist, then there is no fourth dimension of space-time, and the string theory cannot do this most basic necessity of any cosmological theory: revealing that it is incorrect.

QUANTUM CHROMODYNAMICS – AND CHARGE QUANTIZATION

Quantum Chromodynamics is based upon the principles of special relativity and quantum mechanics. It makes predictions that can be tested and used to calculate minute changes in the energy levels of atoms [up to one part per million]. And yet, it cannot explain why all values of electric charges are integral multiples of the electron's charge: a phenomenon known formally as *charge quantization.* In other words, it cannot explain how the most common charge is being generated on the most well-known "Particle" in the world: the lowly electron!

THE STANDARD MODEL – AND GRAVITY; MASS;

Although the standard model appears to be successful in explaining the interactions of particles through the use of the strong, weak, and electromagnetic forces it does not include gravity.

Even more important, the standard model cannot explain why quarks transform into other quarks (change color). And when it comes to mass, the standard model can only state that particles possess mass; it cannot explain *why* they possess mass. The Higgs Boson Particle does not exist.

The standard model is a failed classification of "particles" because some do not exist. It is not an explanation of particles.

Chapter 31
Why All "Particles" Possess the Same Amount of Charge

> Never suspected by 20th Century science, the real reason why all sub-atomic particles possess the same amount of charge is a stunning triumph of the Vortex Theory.

One of the most perplexing phenomenon in the subatomic world is the relationship between charge and particle size and mass. It has long been a mystery why particles of different sizes and masses all possess the same amount of charge.

This problem was considered solved with the proposal of the theory of quantum chromo-dynamics. This theory proposes that the ±1 charges on mesons and barons are created by their quark content. However, this most sophisticated of all theories cannot explain the charges on charged leptons [the electron, muon, and tau]?

For example, how can the anti-particle of the electron – the positron – also possess a charge of +1 without containing any quarks? If quarks are used to explain the charge on the proton, why can't they explain the charge on the positron? Using this logic, if quarks define charge, then the positron and the electron with no quarks should have a charge of 0! Consequently, this failure to explain what is now called the fundamental charge of nature reveals a flaw in this theory!

Fortunately, the answer is easily corrected when it is realized that it is not the quark content that creates the charge of a particle. Instead, it is just the reverse: it is the fundamental charge that dictates what charged quarks the "particle" can possess!

As shocking as it may seem, this Vortex Theory has discovered that all of the charged particles in nature are created out of electrons and positrons! Electrons and positrons can be thought of as containers for other particles. They are in effect "bags" holding the quarks created inside of them!

This idea is not speculation but is based upon the decay of charged leptons, and mesons all of whom want to decay down into lowly electrons and positrons! Revealing that these most basic charged particles in nature CONTAINED quarks that subsequently decayed down into electrons and positrons – the particles that originally contained them!

Chapter 32
The Explanation of the Strong Force

> Reexamining Hideki Yukawa's explanation of the strong force using the principles of the theory proposed at the beginning of this book, it was discovered that it is possible for a virtual particle [a virtual pion] to be passed back and forth between the proton and the neutron. This discovery reveals that Mr. Yukawa's explanation was correct; and that the present explanation using gluons is wrong!

HISTORY

In 1935, the great Japanese Physicist Hideki Yukawa proposed that the strong force could be explained by the existence of a virtual particle approximately 200 times the mass of the electron being passing back and forth between protons and neutrons in the nucleus of the atom.

This particle was designated a "virtual particle" because it passes back and forth so quickly it cannot be seen or detected. At the time of its proposal, this idea was highly speculative because no particle of this mass was known to exist. Then, in 1947, a team of physicists led by Cecil Powell from Bristol University in England discovered the pion. The pion appeared to be exactly the particle predicted by Yukawa: [it felt the strong force, and it was 1/7 the mass of the proton]. However, even though such a particle was discovered to exist in nature, how can it exist within the nucleus; how can it be passed back and forth between protons and neutrons; and how come it moves so fast it cannot be seen?

The answer is found in the newly discovered Vortex Theory's explanation of the Quark Theory.

PROBLEM WITH THE CURRENT VISION OF THE STRONG FORCE

If the Vortex Theory is correct and quarks are indeed small higher dimensional holes existing within a larger 3d hole, then the current belief that gluons are responsible for the strong force is incorrect. The use of gluons and the principles of Quantum Chromodynamics can only explain how quarks are bound together *within* protons and neutrons [3d holes]; they cannot explain what is holding protons and neutrons together.

This paper proposes that the protons and the neutrons within the nucleus of the atom are being held together by the constant exchange of Mr. Yukawa's virtual particle.

HIDEKI YUKAWA'S VIRTUAL PARTICLE

According to the Vortex Theory, when a neutron first approaches a proton, the less dense space surrounding each "particle" distorts the shape of the other. Then, as both distorted holes attempt to straighten out, they move slightly in the direction of the other. As this process continues, with each additional movement increasing the velocity of the particles, accelerating each in the direction of the other until they collide. As they collide, these two holes try to bend into each other, and as they do, some of the space flowing around and around in the neutron's fourth dimensional [4d] torus tries to flow into the fourth dimensional [4d] vortex of the proton.

This movement of space out of the torus causes it to break. As the torus breaks two ends are created: one end connects to the proton's 4d vortex; the other end emerges back into 3d space, turning it into a proton. At the same instant, the broken end of the original proton's 4d vortex now wraps around and envelops *it*, changing its vortex into a torus; changing its identity - "metamorphosing" the proton into a neutron; (and the neutron into a proton). However, just as soon as the switch occurs, the process instantly begins all over again, causing the newly formed "particles" to revert to their former identities. This constant switching of identities keeps one "particle" pressed tightly against the other - becoming the strong force of nature.

Figure 32.1

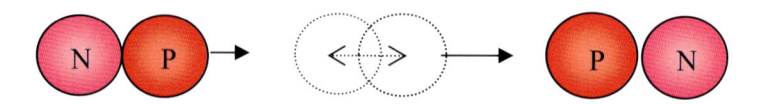

[Note: because the extra three dimensional space surrounding the larger neutron does not have time to "deflate," but instead is transferred to the proton beside it, no anti-neutrino particle is allowed to form as it does when the neutron is alone in free space. Also, because the extra 3d space surrounding it does not have the opportunity to "deflate", the end of the vortex that is being passed back and forth is much larger than the end that would normally shrink in size to form the electron.]

THE CREATION OF YUKAWA'S VIRTUAL PARTICLE…

Even though the creation of these continual and seemingly instantaneous microscopic metamorphoses is a fascinating and dramatic event, an even more remarkable event is taking place within the proton and neutron. As the proton and the neutron constantly merge together, change identities, and split apart, two quarks are continually being passed back and forth between them.

As the two particles merge together, an up quark from the proton is passed to the neutron, and a down quark from the neutron is passed to the proton. And as this transference occurs, as the two different quarks pass by each other on their continuous journey to the opposite "particle," for a brief instant a virtual pion is created!

In Figure 32.2 below, the proton and neutron are touching; in Figure 32.3, as the proton and neutron begin to change identities, an up quark from the proton and a down quark from the neutron approach each other; in Figure 32.4, as the proton and the neutron merge together, the up and down quarks pass each other creating for a brief instant a **Virtual Pion (with 1/7 the mass of the proton)**; and finally, in Figure 32.5, after the transformation is complete, the neutron and proton have switched identities.

Figure 32.2   **Figure 32.3**

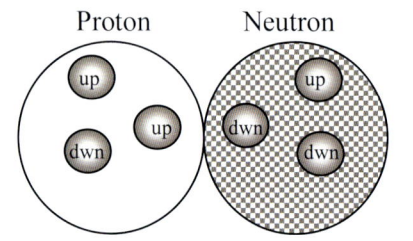

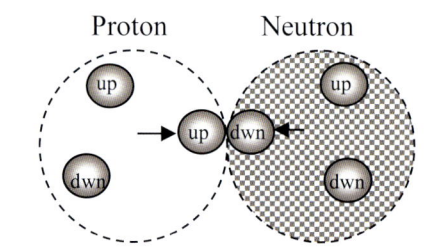

Figure 32.4   **Figure 32.5**

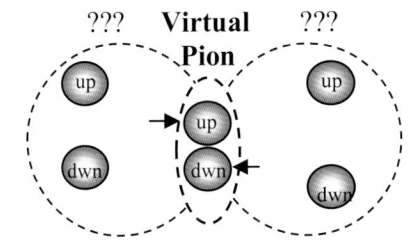

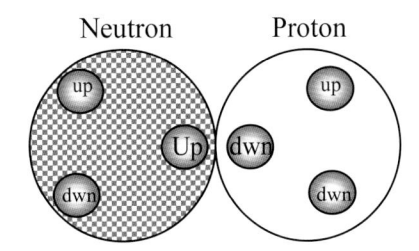

If the Vortex Theory is true, then Hideki Yukawa's proposal of a virtual particle being passed back and forth between the proton and the neutron is correct, and he needs to be posthumously recognized for this great achievement.

ALPHA PARTICLES...

"Cooper pairs" of protons and neutrons create an alpha particle construction of the nucleus...

The key to understanding why the nucleus is constructed out of alpha particles is the alpha particle's *neutral spin*. This neutral value reveals that the spins of all four nucleons cancel. And if the Vortex Theory it true, and the vortices exist, because the spin states of the two electrons in a helium atom are opposites, the spin states of the two protons are also opposites; and to complete the neutrality, the spin states of the neutrons must cancel too. It is this combined neutrality that appears to be holding the alpha particle together:

As mentioned previously, the magnetic moments of "particles" is created by the rotation of the space around them; [fermions are spinning around a 4d axis]. Opposite spin states create opposite rotations creating opposite polarities. Because the 2 protons and the 2 neutrons in the alpha particle possess opposite spins they possess opposite polarities. These opposite polarities cause the 2 protons and the 2 neutrons to be attracted to each other much like *Cooper pairs*; binding one proton neutron pair to the other – binding the alpha particle together.

When researching deuterium nuclei, it is interesting to note their spin state is 1. This reveals that the spin state of both the proton and the neutron are the same. If random chance was involved in the creation of the deuterium nucleus, the probable odds that such a condition could occur naturally would be astronomical. Consequently, there has to be a reason why the spin states of both the proton and the neutron are the same:

According to the Vortex Theory the proton and neutron are constantly switching identities. Although it is conceivable that the two nucleons could switch identities while having different spin states it appears to be highly unlikely. The vortex would constantly be twisted out of shape and then twisted back into shape. When twisted out of shape, it would go into a higher energy state. It would need energy to make the first switch and would release energy when it switched back; and would then have to acquire additional energy to repeat the process.

Also, it is possible for a proton and neutron of different spin states to switch identities if one is being constantly "turned upside down" in 4d space. [Note: in 3d space, if one of two spinning wheels is turned upside down, from its perspective, it is now spinning in the opposite direction to the other.] However, and again, this condition appears to be highly unlikely. Because for one of the Nucleons to turn upside down in 4d space would require the constant switch from a higher to a lower energy state.

Other evidence is found in ***quark confinement***. If the Vortex Theory is true, and quarks [4d holes] are indeed sheathed within larger 3d holes, then the present understanding of the role played by gluons in creating the strong force is incorrect. Gluons could be responsible for holding 4d holes together but <u>cannot</u> be used to explain what is holding the larger 3d holes together.

Stated more simply, if quarks were baseballs within basketballs, it is easy to see that gluons within the basketballs could hold the baseballs together, but because they are inside, they cannot hold the larger basketballs together.

And finally, since these larger 3d holes surround the volume of 4d space within, it is extremely important to understand that this 3d hole is the surface of this 4d volume of space within it. Because a volume is confined within its surface and cannot leave its surface, *it is the 3d hole that is responsible for quark confinement –* <u>not</u> the strong force! It is the geometry of 3d space itself that is keeping the 4d holes we call quarks confined. <u>*Therefore, the belief that quark confinement is created by a strong force that increases as distance increases is also a mistake!*</u>

Chapter 33
The Alpha and Beta Particles

> The release of an electron from within the nucleus of an atom is called Beta Decay. This release is accompanied by the change of a neutron within the nucleus to a proton.
>
> The explanation is simple. The 4d torus creating the neutron breaks; one end is the proton, the other is the electron. The proton stays within the nucleus while the electron is thrown free.

THE EXPLANATION OF THE ALPHA PARTICLE

One of the great observation achievements of early particle scientists was the discovery that only three types of particles are expelled from the nucleus of an atom: alpha particles, beta particles, and gamma rays.

Now the gamma rays don't present a problem because they are nothing more than energetic photons; nor do the beta particles because they are merely electrons. However, the alpha particles –which are nothing more than helium nuclei – do present a very special problem: why only alpha particles? Why are only alpha particles thrown out of the nucleus? Alpha particles consist of two protons and two neutrons. So, what is so special about this arrangement? Why don't we see particles made up of one proton and one neutron, or three protons and three neutrons, or four protons and four neutrons, or more?

The answer is found in the true nature of the strong force. The strong force is not an attraction between particles, but rather the continual transformation of a proton into a neutron and a neutron back into a proton. This process is easy to observe between particles that are paired with each other, but it is also easy to observe in two pairs of particles.

Although already explained before, when a proton-neutron pair is close to another proton-neutron pair, the continual transformation keeps all four particles "stuck" together:

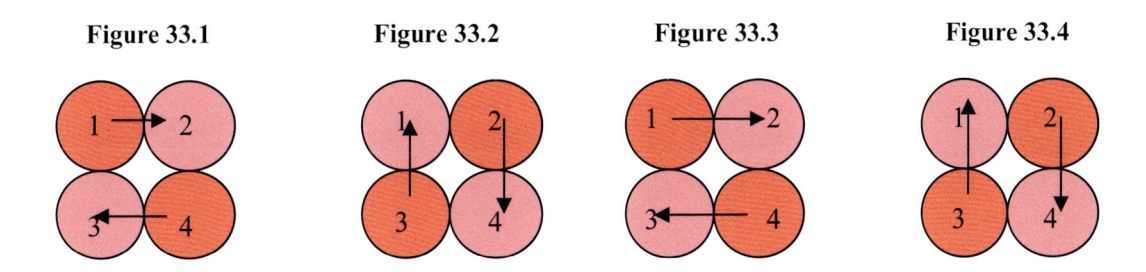

Figure 33.1   **Figure 33.2**   **Figure 33.3**   **Figure 33.4**

[note: red = proton: lavender = neutron]

In the above sequence note how 1 becomes 2, 2 becomes 4, 4 becomes 3, and 3 becomes 1 again, completing the circuit. Note also, that if 1 became 2 as 4 became 3, and then 2 became 1 again as 3 became 4 again, each pair would be acting as a separate set instead of one set. In this circumstance, set 1&2 and 3&4 would be separate from each other and would not be held together by the strong force. Instead, the two sets would be pressed together inside of the nucleus due to the bent inward regions of space that surround each particle.

Because heavy nuclei are made out of many protons and neutrons it is possible that a majority of the particles form these structures: making the interior of the nucleus consist mainly of alpha particles.

It is speculation that the spontaneous emission of alpha particles from the nucleus of a heavy atom is a result of the many extra neutrons trying to swap with the protons in the alpha particles causing them to disengage, to suddenly move in an opposite direction from these other neutrons; causing the alpha particle to be ejected out of the nucleus.

THE INSTABILITY OF ISOTOPES

The excess neutrons that create isotopes of a particular atom are unstable because the switching process that takes place back and forth between protons and neutrons must now encompass more neutrons. The roundtrip that usually takes place between two protons and two neutrons must now take place between one or more neutrons. This creates an unstable situation because a neutron is allowed to remain a neutron for a longer period of time. This allows the space within the neutron to circulate for a longer period of time allowing the torus that creates the neutron to decay faster. Allowing the neutron to decay into a proton and electron [and anti-neutrino] ejecting the electron out of the nucleus.

Chapter 34
The Explanation of the Double Slit Interference Patterns Created by Electrons!

> The mystery of how one electron appears to go through the two slits at the same time and creates an interference pattern is explained by the large dense region of space that surrounds the electron - preceding the electron through the two slits, creating the interference pattern in the process.

The great mystery of how a single electron can create the wave like interference pattern in Thomas Young's famous twin slit experiment is easily explained by the Vortex Theory.

The wave-like interference pattern made by projecting either light or matter through the twin slits *is created by the dense regions of three dimensional space surrounding electrons.* In comparison to the size of the electron, the dense region of space that surrounds it is massive.

Figure 34.1

Note: in relation to the size of the electron, the region of dense space that surrounds the electron is so massive it is impossible to draw the proportionate sizes.

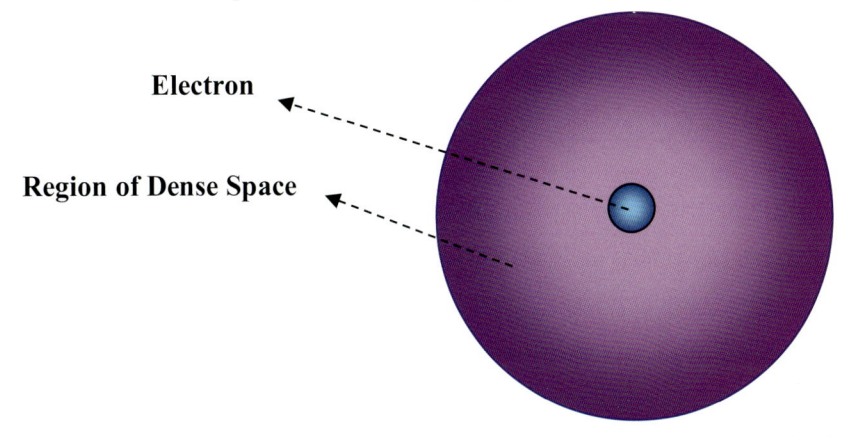

Electron

Region of Dense Space

This region is so massive, that when an electron approaches the twin slits, its denser region arrives first.

Figure 34.2

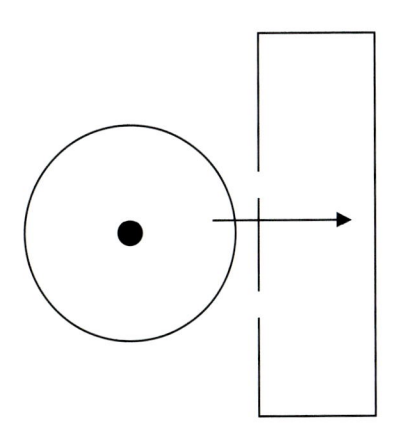

Because the dense region arrives first, it begins to go through the slits before the electron.

Figure 34.3

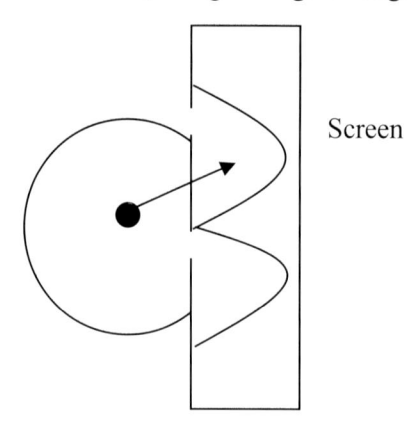

As this denser region moves through both slits simultaneously, the two waves interfere with each other and begin to create the invisible wave patterns of dense and less dense space on the screen.

Figure 34.4

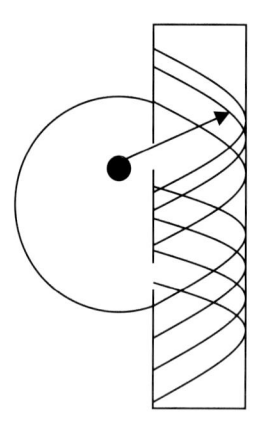

When the electron finally moves through one of the slits, it moves into a region of space that is alternately dense, less dense, dense, less dense, etc. Consequently, the surface of the electron is distorted towards *one of the less dense regions*, avoids the denser regions, and travels in the direction of this less dense region, striking the screen.

Figure 34.5   **Figure 34.6** 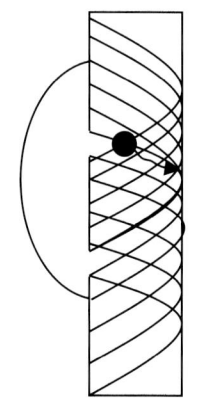   **Figure 34.7**

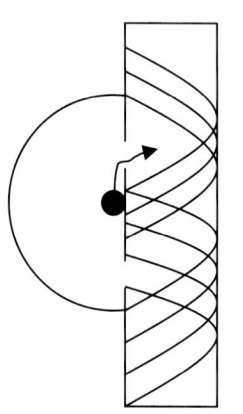

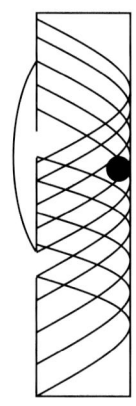

Note, in Figure 34.6, as the electron passes though one of the twin slits, note how the dense region behind the slit has not yet passed all of the way through. This follow through of the sphere of dense space behind the electron continues the interference pattern until the electron strikes the screen. Note too, this picture shows the electron going through the top slit. If the region of dense space were oriented more towards the bottom slit, it would have gone through that one.

When this same situation happens over and over again, it is their collective statistical behavior that sends the electrons to the less dense regions, avoiding the denser regions, and creates the light and dark strips seen on the screen.

Figure 34.8

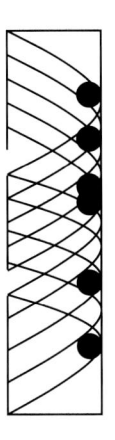

Because it can now be seen how a single electron can create the interference pattern, some important misunderstandings can now be laid to rest.

#1. Perhaps the most important misunderstanding is the mistaken idea that somehow the electron was thought to go through both slits at the same time. As can now be seen, this is not what is happening.

#2. Equally important is the idea that photons of light are somehow interfering with each other, "canceling each other out" creating the dark spots on the screen. This is also a mistaken idea. Instead, it can now be seen how electrons avoid the denser regions and are attracted to the less dense regions. [Note, the three dimensional hole of the electron is distorted in the direction of the less dense regions of space, causing it to move in those directions.]

Chapter 35
How "Looking" at the Electron in the Two Slit Experiment Changes the Results?

> One of the fantastic phenomenon of subatomic physics is the seemingly mystical fact that just to look at the electron two slit experiment somehow changes it!
>
> In the past, this bizarre fact conjures up explanations that range from the paranormal to the spiritual. However, with the discovery of the Vortex Theory, the phenomenon is now easily explained.

As explained in *The Vortex Theory*, both the electron and the photon are surrounded by regions of space bent outward. In the photon two slit experiment, incoming photons do not disturb the "flight" of the photons involved in the experiment. Like two waves approaching each other upon the surface of the ocean that pass right through each other, photons do the exact same thing [unless they are gamma rays that hit and create electron positron pairs]. But not electrons!

Because an electron is a hole in space, an approaching photon cannot pass through it. Instead, the bent outward region of the photon and the bent outward region of the electron push against each other and move the electron, changing its trajectory.

The eye does this too.

The eye not only absorbs light but reflects it too. Consequently, when we look at an object, light reflected off of the eye is "aimed" directly back at the object we are looking at. So, when we look at the electron double slit experiment the light coming off of our eyes hits the electrons in the experiment, changing their trajectory creating changes in the experiment.

In the figure below, the photon reflects off of the eye, hits the bent outward space surrounding the electron, and changes its trajectory: [note that the photon, like the electron, is surrounded by a *dense region of 3d space*].

Figure 35.1

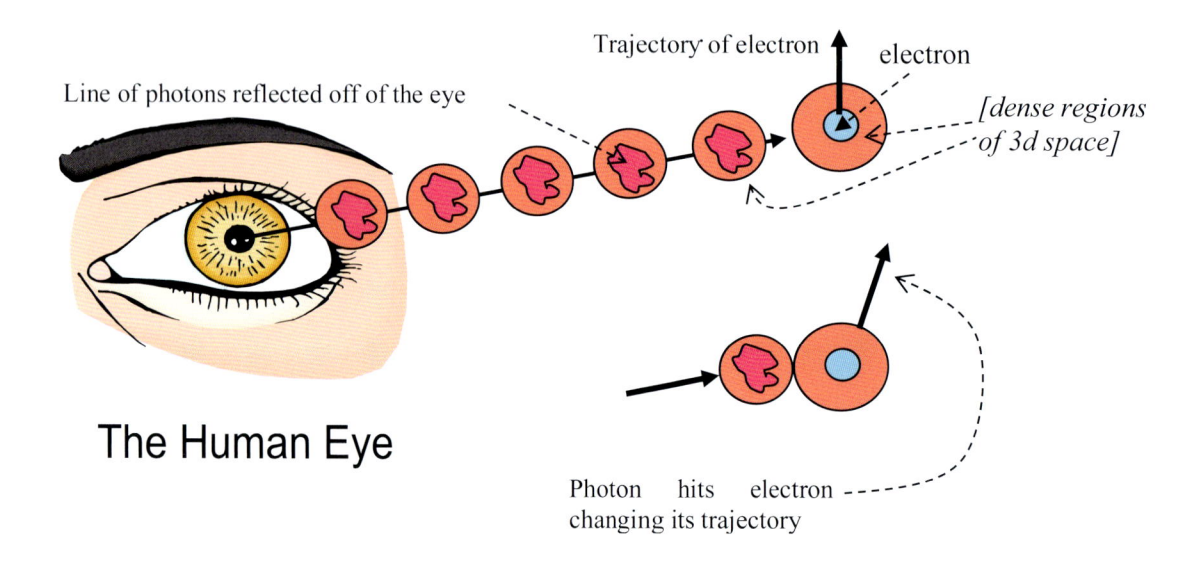

Chapter 36
Why the Electron Does Not Like to Be Confined

> Another mystery that can be cleared up by the Vortex Theory is the strange phenomenon regarding the failure to confine the electron.

When electrons are confined within a small "box," the electron eventually escapes. The reason why this happens comes from the realization that the electron is the other end of a vortex of flowing space.

This other end is either a free positron or a free proton. When this free particle moves, the movement is transferred through the vortex to the electron. This causes the electron to tunnel past the wall of the box by dipping down into 4d space, then back upward into 3d space.

PART III
PARTICLE DECAYS AND COLLISIONS

Chapter 37
Decay of the Neutron

> The theory at the beginning of this book that the universe is turning inside out, is no better illustrated than in the "particle" science calls the Neutron. According to this Vortex Theory, the neutron is the combination of a proton and an electron: a hole within a hole. As such, it is created by space turned inside out – further proof of the theory that space is one gigantic particle turning itself inside out. But this is not all. The decay of the neutron also explains how all other "zero" charged particles decay.

The discovery that the neutron is a 3d hole bent into 4d space, surrounded with a 3d hole bent outward into 4d space is discussed in *The End of the Concept of Time PART 2 – The Vortex Theory of Atomic Particles*. However, why this structure should then decay was not understood until years later. The explanation is presented here.

THE DECAY OF THE NEUTRON

One of the great mysteries of particle science is the decay of the neutron. Why does the neutron decay in approximately 10.5 minutes when it is outside of the nucleus of an atom?

The answer is found in the quark content of the neutron, their alignment, and their rotation. For example, inside the neutron are two down quarks and one up quark:

Figure 37.1

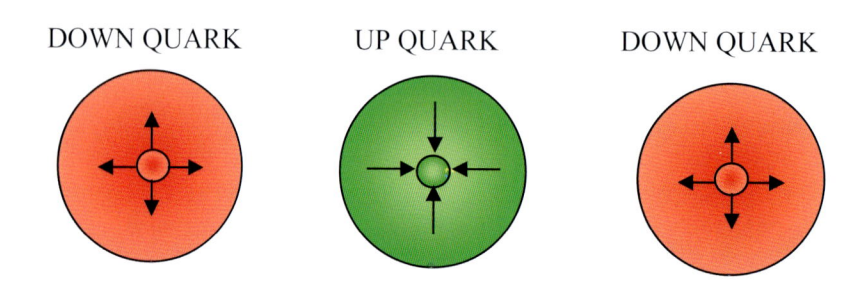

DOWN QUARK   UP QUARK   DOWN QUARK

Notice in the above drawing, how 4d space [these are 4d holes] is flowing into the up quark and out of the down quarks. The outward flow of space from the two down quarks into the up quark causes both down quarks to align on either side of the up quark as seen above; and in the drawing below:

Figure 37.2

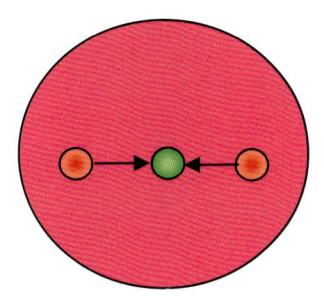

In the figure above, notice how the neutron's surface is spherical. However, as the two down quarks rotate about the up quark, this situation changes. A harmonic wave begins to be generated upon the surface of the neutron as seen in the figure below:

A harmonic wave distorts the surface of the torus, remember, the neutron is a fourth dimensional torus:

Figure 37.3

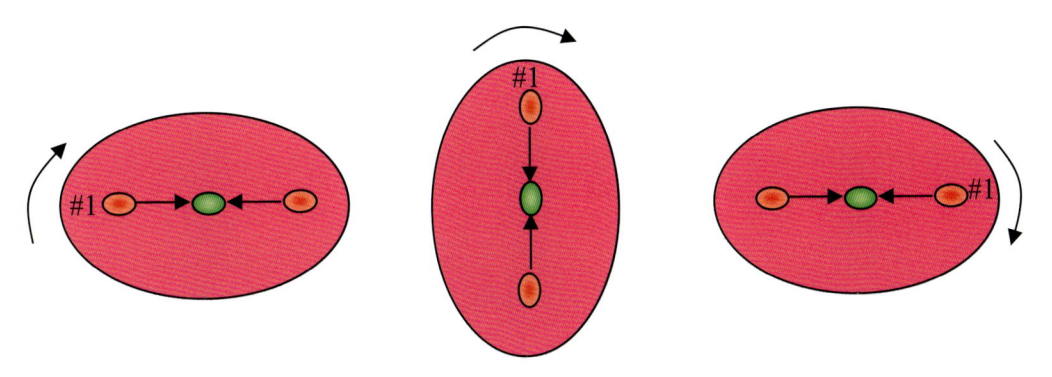

In the above drawing, notice how the rotation of the down quarks causes the surface to distort. As they continue to rotate a harmonic is set up upon the torus which is the surface of the neutron:

Figure 37.4

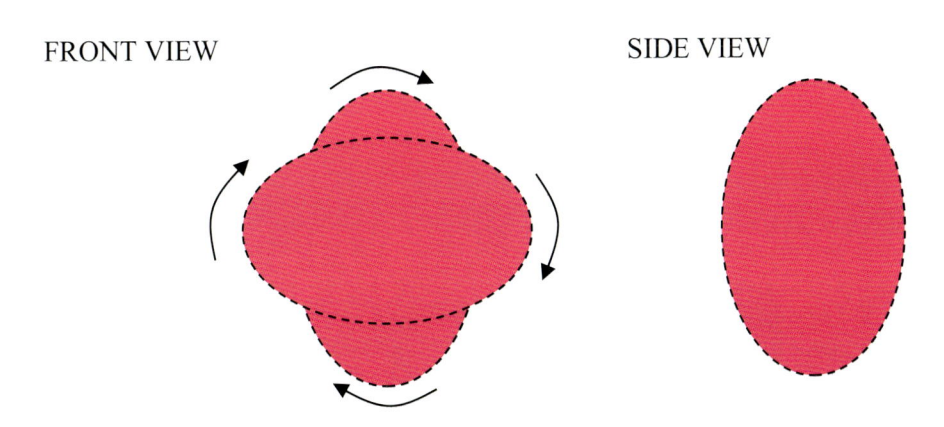

FRONT VIEW   SIDE VIEW

As the rotations continue, the distortions elongate the neutron; increase the amplitude of the harmonic wave, and the frequency of the resonance. The distortion of the torus allows the rotational speed of the quarks to increase, increasing the resonance even more:

Figure 37.5

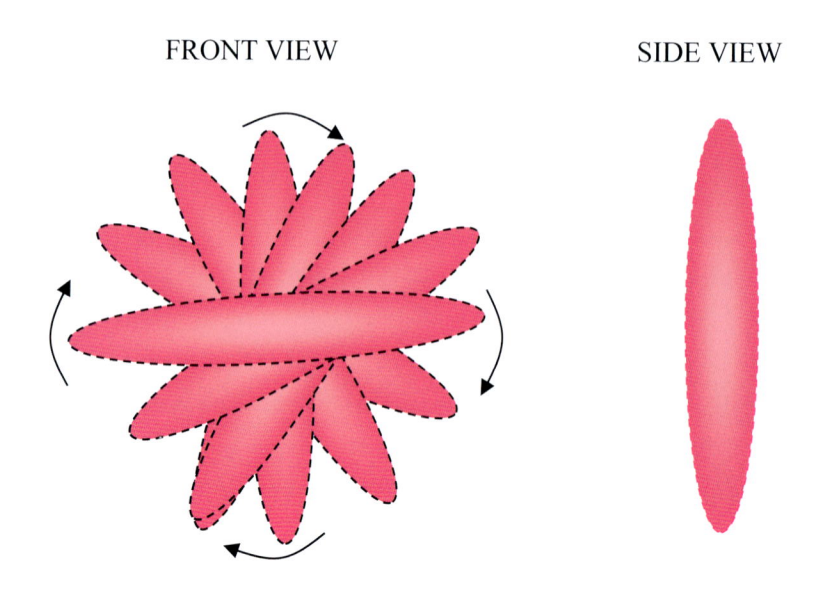

FRONT VIEW   SIDE VIEW

The amplitude of the distortion of the surface continues to increase until finally the elastic modulus of the space is surpassed and the torus breaks:

Figure 37.6

NEUTRON IN FREE SPACE

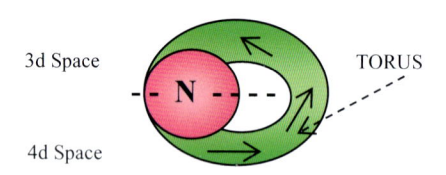

Vortex breaks

Figure 37.7

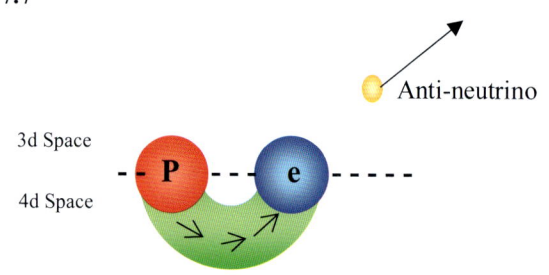

Note how the vortex breaks revealing one end to be the proton, the other to be the electron. The anti-neutrino is the extra volume of space that was trapped within the neutron's volume.

Figure 37.8

As the proton and the electron move away from each other, the 3d space again begins to flow into the proton and out of the electron.

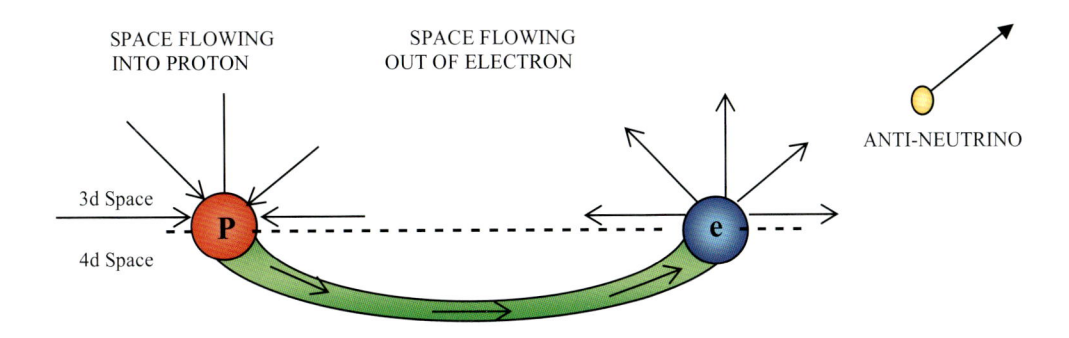

However, expanding the theory that the universe is created out of one single particle turning inside out, [creating two volumes of space with two different elasticizes], it is now easy to see that the creation of the same fundamental charge of particles throughout the universe is a result of the maximum amount of space from side 1 that can flow into a hole whose size is a function of the elasticity of side 2: the contracting volume.

If more space tries to flow into the hole it cannot because the elastic modulus of Side 2 will only allow an opening of a certain size: where the pressure of the flow pushing outward from side 1 equals the inward pressure of the space from side 2 that is trying to close the hole.

The same equilibrium will occur even if the density of space changes for larger particles, such as proposed in Book 1. Note: larger, more massive particles are nothing more than enlarged electrons and positrons containing quarks [even muons and tau particles as will be explained later]. As a consequence, the present idea that quarks create the charges upon the particles they inhabit is a mistake.

Chapter 38
New Particle? The Tunneling Pion!

Using the principles of the Vortex Theory, it was discovered that when the neutrino collides with the neutron, or when the anti-neutrino collides with the proton, "tunneling" quarks are released: a neutron releases the down, anti-up quarks; the proton releases the up, anti-down quarks. The down, anti-up quarks create a tunneling negative pion, while the up, anti-down quarks create a tunneling positive pion. (When the energy is high enough these pions do not tunnel, but instead become the pions seen during the decays of nucleons.) Both particles travel (tunnel) through fourth dimensional space just beneath the surface of three dimensional space. These two particles are the precursor of a new classification of particle whose mass and energy characteristics are undetectable until colliding with another particle.

APPLICATION OF THE VORTEX THEORY TO THE QUARK THEORY

When the principles of the Vortex Theory were applied to the Quark Theory, many new and exciting discoveries were made. Among them were the two sides of space; and that ±1/3 charged quarks are formed on one side (SIDE 1), while ±2/3 charged quarks are formed on the other side (SIDE 2). Also, when a quark decays on one side of space, it pulls or pushes the other side inward or outward creating quarks on the other side: explaining how quarks change "Flavor." This explanation has led to many other breakthrough discoveries, such as, the explanation of how a neutron is changed into a proton when struck by a neutrino; and how a proton is turned into a neutron when struck by an anti-neutrino.

These two discoveries have an added benefit because they reveal that specific quarks are created and decay with specific baryons. For example, when the neutron is hit by an electron neutrino, a down quark is not changed into an up quark. Instead, one of the down quarks on Side 1 decays and "tunnels". As it does, it pushes Side 2 outward creating the up, anti-up pair of quarks. The up quark stays within the newly formed proton while the anti-up quark tunnels alongside the down quark. (How this happens will be explained in the DISCUSSION.)

The down and the anti-up quark form a tunneling negative pion. The presence of this particle is undetectable because it travels within 4d space at the speed of the space flowing in the vortex (the speed of light) just below the surface of 3d space. The collisions with other particles and the resultant decay products are discussed in DETECTING THE TUNNELING NEGATIVE PION.

When a proton is hit by an anti-neutrino, it has a specific number of quarks associated with it too. It has the up, up, down of the original proton, and a newly formed down, anti-down pair of quarks (five quarks in all). The down, anti-down quarks are formed when the up on Side 2 decays and tunnels. When it decays, it pulls Side 1 outward creating the down, anti-down pair of quarks. The up, down, down, quarks form the neutron, while the other up and the anti-down tunnel.

The up and the anti-down quark form a tunneling positive pion. The presence of this particle is undetectable because it travels at the speed of the space flowing in the vortex (the speed of light) just below the surface of 3d space. The collisions with other particles and the resultant decay products are discussed in DETECTING THE TUNNELING POSITIVE PION.

DISCUSSION

The explanation of the creation of the tunneling pions first begins with the creation of the proton and the neutron. Although 20th Century science sees Figure 38.1, according to the Vortex Theory model of the universe, what is really happening is seen in Figure 38.2:

Figure 38.1. The 20th Century model traditional view:

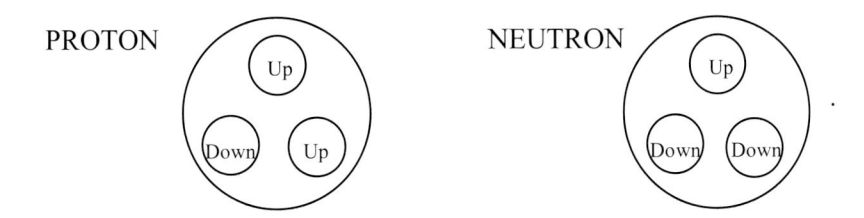

In the 20th Century model of the proton and neutron, quarks are all "particles" existing somehow within the "particles" called protons and neutrons. In this model, neither the charges of quarks nor the charges of the "particles" they inhabit are explained. Neither is it known how quarks change flavor:

Figure 38.2 The Vortex Theory model: in the vortex model of the atom, all that could not be explained by the 20th Century model can now be explained.

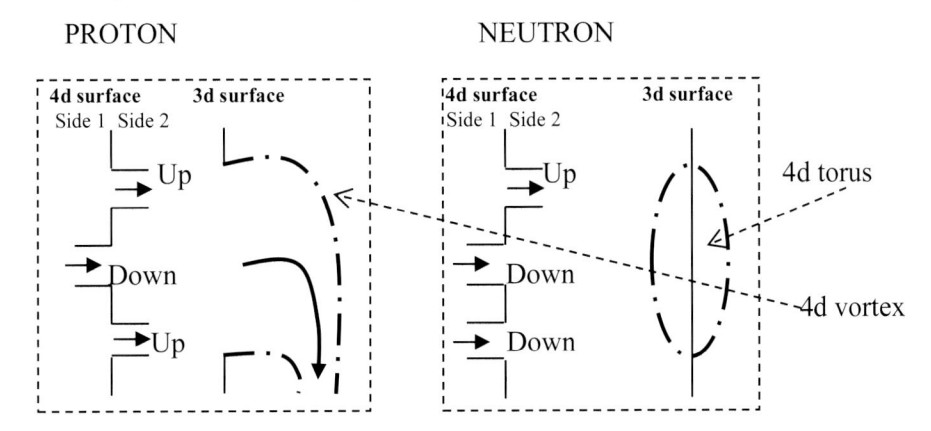

In Figure 38.2, notice how the up quarks are formed on Side 2 of space while the down quark is formed on Side 1. The quarks are formed on the fourth dimension of space (4d space). The arrow inside the lines represents *one end* of a fifth dimensional vortex of flowing space. The other end of this vortex is connected to the quark's anti-particle somewhere else in the universe.

In the above schematic drawings, the arrows in the up quarks point away from the 4d surface and represent the positive charges of the up quarks; the down quark's arrow points toward the 4d surface and represents its negative charge.

The opening between the dashed-dotted curved lines represent the three dimensional hole that encases the quarks. The dashed lines in the proton represents the three dimensional space (3d space), that flows into 4d space and encircles the fourth dimensional vortex. The arrow flows away from the 3d surface and represents the positive charge of the proton.

In the neutron, the same representations apply, with one exception: the oval dash-dotted line represents the neutron's fourth dimensional torus: the 4d vortex has broken and formed a 4d torus creating the neutron's neutral charge.

116

Using the above model, we observe the collision between a proton and an electron anti-neutrino:

Note: in the Vortex Theory model of the universe, the electron neutrino is a quantized three dimensional transverse wave [similar to a ½ sphere] traveling upon the surface of 4d space.

Figure 38.3 *The collision between a proton and an anti-neutrino*:

In the drawings below, the anti-neutrino approaches and then strikes the proton:

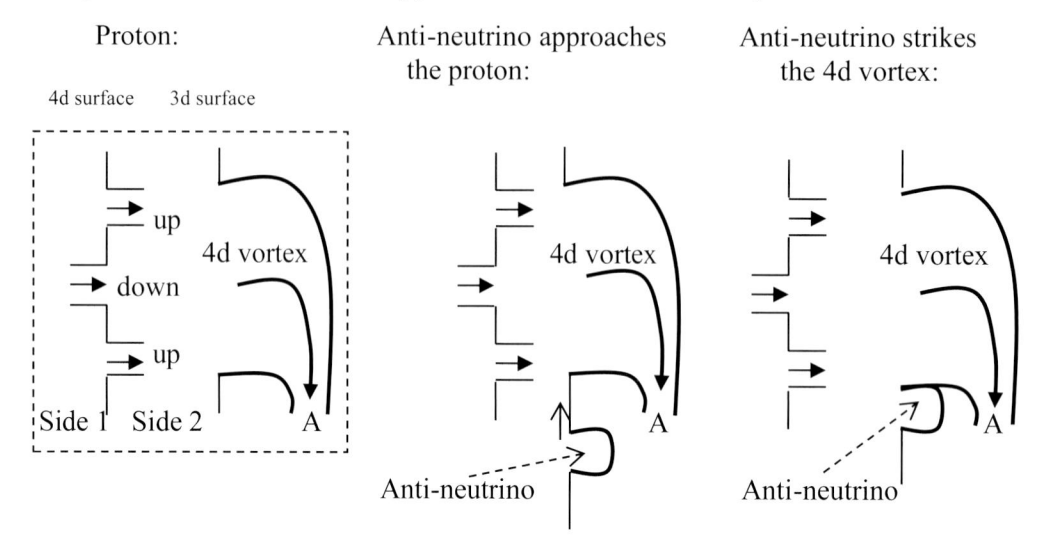

Moving at the speed of light, when the anti-neutrino strikes the 4d vortex, *its volume of space bent into side 2 is suddenly added* to the 4d vortex's volume of flowing space that is also bent into side 2. This almost instantaneous inflation causes three simultaneous events to occur:

#1. The 4d vortex breaks. One end curls back upon itself creating a 4d torus; while the other end continues to flow into side 2 becoming the positron:

Figure 38.4

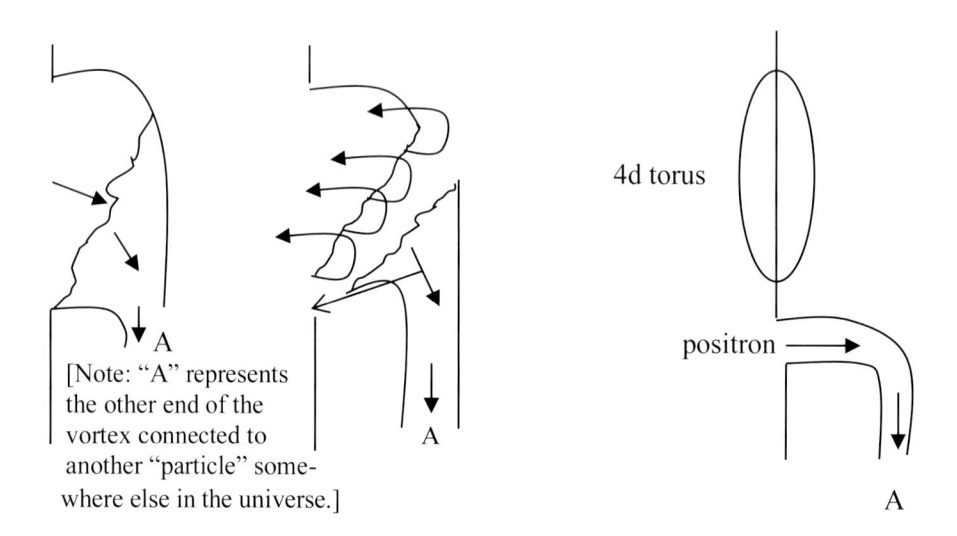

[Note: "A" represents the other end of the vortex connected to another "particle" somewhere else in the universe.]

As the electron anti- neutrino collides with the proton, its volume of 4d space bent into side 2 suddenly adds to the volume of 4d space that is bent into side two. This almost instantaneous increase in the volume of the space within the proton creates the effect of a down quark [down quarks add 4d space to the 4d volume]. This causes the up quark nearest to the collision to decay. The decay pulls side 1 outward creating the down, anti-down vortex combination: the down quark stays, the anti-down quark tunnels.

As the down, anti-down quarks are being created, and the breaking 4d vortex forms a torus, and a circulating flow within the newly forming neutron begins; also, the anti-down and the decaying up quark tunnel off into space, side by side away from the site of the collision.

[Note: it appears as if the anti-down quark and the up quark should unite to form a positive pion inside the vortex created by the positron. However, there is not a large enough volume of 4d space present. If there were, the increased volume would allow the positron's vortex to increase in size, allowing the anti-down and the up quark to remain within it creating a positive pion. However, this effect is indeed created during the decay of nucleons.

#3. When the up quark deflated, creating the down, anti-down quark pair, the volume of space within the neutron is suddenly filled up by the presence of the down quark. Consequently, the outward flowing space at the other end of the down vortex, [the anti-down quark], has no place to go. It cannot stay within the neutron, so upon its formation, the anti-down quark immediately deflates and tunnels through 4d space. As it deflates, its inward flowing space causes side 1 to pull inward, creating the effect of a muon neutrino. However, because the outward flowing space of the deflating up quark pulls side 2 inward, creating the effect of a muon anti-neutrino, as seen from the 3d surface, *the two opposite bends in space cancel each other and as the two quarks tunnel off together, no neutrino is seen, all that is left is the neutron and the positron:*

(Note: if there is enough volume in the positron, the two tunneling quarks will enter the positron and form the positive pion.)

Figure 38.5   **Figure 38.6**

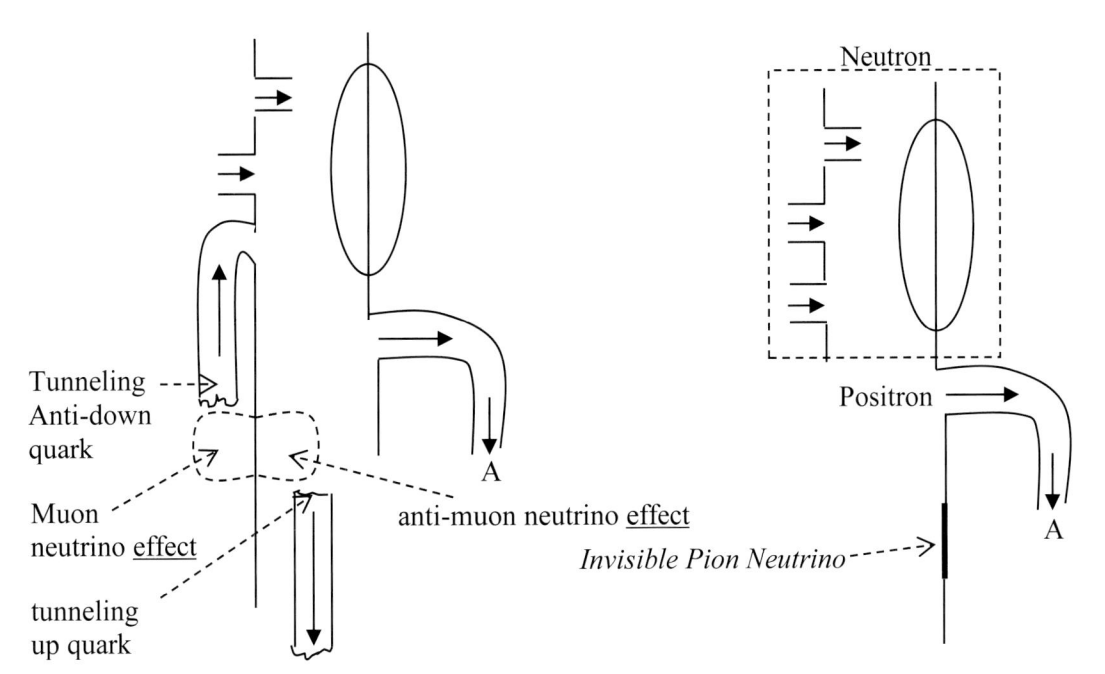

(The muon neutrino is explained in Chapter 45 on page 140)

Note, because the tunneling anti-down quark and the tunneling up quark travel side by side, their simultaneous opposite pulls upon the opposite surfaces of space cancel, and no neutrino or anti-neutrino is seen: in effect, both the matter and energy effects of the tunneling pion are invisible.

The collision between the electron neutrino and the neutron:

As the electron neutrino collides with the neutron, its volume of 4d space bent into side 1 suddenly subtracts from the neutron's internal 4d volume of space that is bent into side 2. This almost instantaneous decrease in the volume of the space within the neutron creates the effect of an up quark [up quarks remove 4d space from the 4d volume]. This causes the down quark nearest to the collision to decay. The decay pushes side 2 outward creating the up, anti-up combination: the up quark stays while the anti-up tunnels.

Figure 38.7

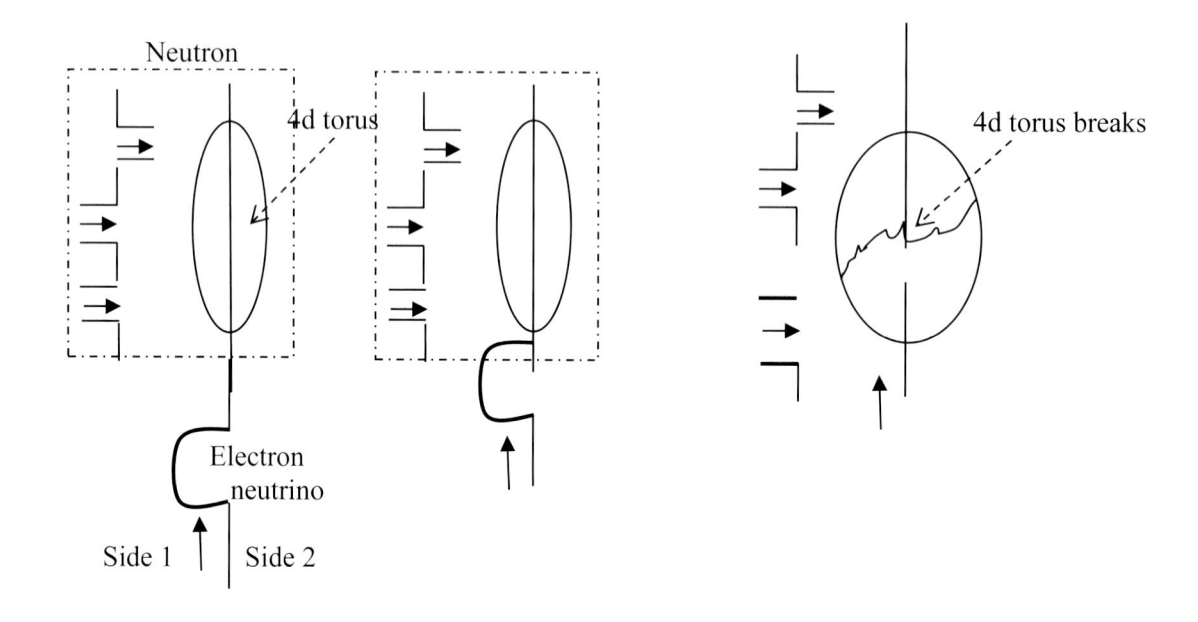

As the torus breaks and begins to reform the two ends of the 4d vortex, because the outward flowing space at the other end of the up vortex, [the anti-up quark], adds to the volume of the proton, it cannot stay within the reforming proton. Instead, it immediately deflates and tunnels through 4d space. As it deflates, its outward flowing space causes side 1 to push outward, creating the <u>effect</u> of a muon neutrino. However, and at the same time, because the down quark has also deflated, its outward push into side 2 creates the <u>effect</u> of a muon anti-neutrino. These two opposite effects cancel each other, and no neutrino is seen. All that is left is the proton and the electron seen below:

(Note, the breaking 4d torus creates the W⁻ Particle. *(W particles are collapsing 4d torus.)*

Figure 38.8 As the torus breaks…

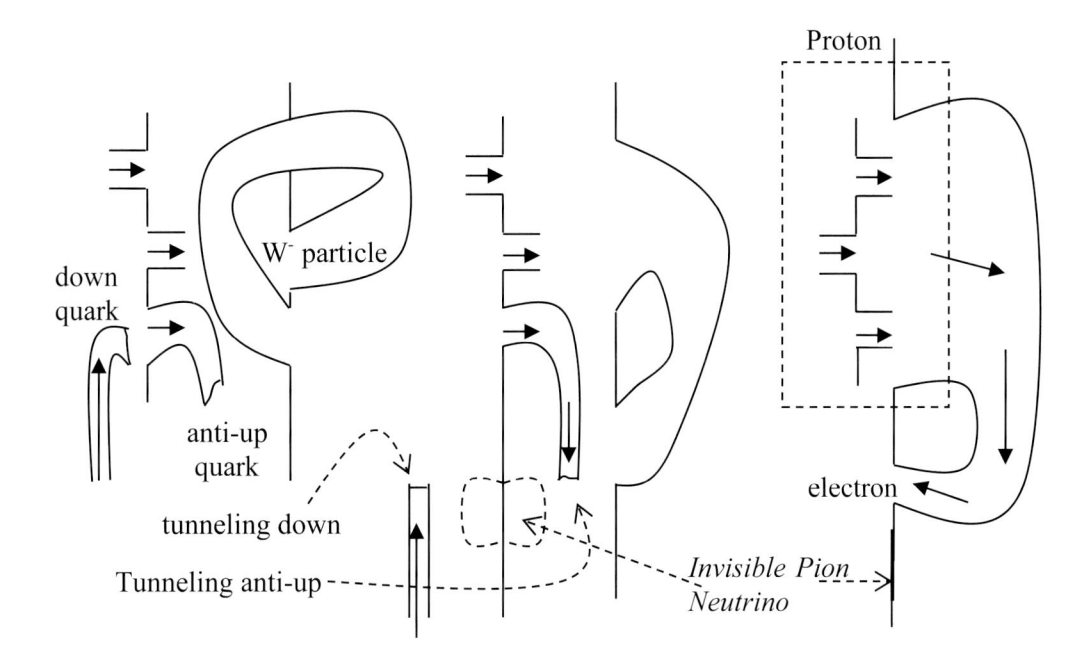

As before, because the tunneling down quark and the tunneling anti-up quark travel side by side their two simultaneous but opposite pressures "push" against each other upon the two different surfaces of space and cancel each other and no neutrino or anti-neutrino is created.

DETECTING THE PRESENCE OF THE TUNNELING NEGATIVE PION.

The detection of the tunneling negative pion consisting of tunneling down and anti-up quarks is seen in its decay products during collisions with other particles. When the tunneling negative pion collides with a proton, because the tunneling negative pion is not sheathed in a fourth dimensional vortex as the proton, the decay products are limited to particles created within the breakup of the proton's vortex. For example:

A. Collisions with protons:

#1 if the positive fourth dimensional vortex tries to break off to form a neutral particle and another positive particle, it does not have the correct quark content. When the up, up, down of the proton is added to the down, anti-up of the tunneling negative pion there is no way to create just a neutral and positive particle from the five quarks. Even if the up, anti-up annihilate and push side 1 inward to create a down, anti-down pair, there still is not the proper quark content.

#2 if a section of the positive vortex breaks off and the ends join to form a neutral particle, then the quark content allows the formation of a neutral particle, a positive particle, and a neutrino. Hence, during the collision with a proton, if the momentum of the accelerated proton is high enough, then the up, down, down form a *neutron*; the other up forms the *positive muon*; and the tunneling anti-up forms a *muon neutrino*. This reaction obeys the conservation of charge [remember the word *negative* in the tunneling negative pion has no 4d vortex, hence, the tunneling pion has no charge; the word *negative* is only a designation used to distinguish it from the *positive* tunneling pion]; the conservation of baryon; and the conservation of lepton laws.

#3 although there is the proper quark content to create the neutron and a positive and negative muon, the vortex cannot break into the proper sections to create this combination of particles [it would violate the law of the conservation of charge].

B. Collisions with neutrons:

If the tunneling negative pion strikes the neutron, the breakup of the neutron loop limits the resultant particles that can be created. For example:

#1 if the tunneling negative pion strikes the neutron the quark combination will consist of an up, down, down, down, anti-up. Because the neutron vortex is caught in a loop, it can only break exposing its negative and positive ends or break again creating another loop limiting the possible particles that can be created.

When the vortex loop breaks, if the momentum of the neutron is high enough, the particles that can be created are the *omega minus* particle with its down, down, down quarks; the *positive muon* with its up quark; and the tunneling anti-up forms the *muon neutrino*. Again, this reaction obeys the conservation of charge, conservation of baryon, and the conservation of lepton laws.

C. Collisions with electrons:

If the tunneling negative pion strikes a low energy electron it will create a negative pion.

THE TUNNELING NEGATIVE PION IN CONCLUSION

When neutrons are struck by neutrinos, it is theorized that the tunneling negative pion will be produced.

The detection of this tunneling negative pion will be discovered in three types of collisions: by the collision with low energy protons producing a neutron, a positive muon, and a muon neutrino; by the collision with a low energy neutron creating the omega minus, the positive muon, and the muon neutrino; or by the collision with a low energy electron creating a negative pion.

If the protons are accelerated to high energies, then the anti-up quark in the tunneling negative pion can annihilate with a high energy up quark in the accelerated proton and bend side 1 of space outward to a greater degree creating an additional bend in 5d space creating the strange, anti-strange pair instead of the down, anti-down pair. This will then result in a cascade of particles created by the decay of the strange and anti-strange quarks, making it harder to detect the presence of the tunneling negative pion.

DETECTING THE PRESENCE OF THE TUNNELING POSITIVE PION

Just like the tunneling negative pion, the detection of the tunneling positive pion consisting of tunneling up and anti-down quarks is seen in its decay products during collisions with other particles. When the tunneling positive pion collides with a proton, because the tunneling positive pion is not sheathed in a fourth dimensional vortex as the proton, the decay products are limited to particles created within the breakup of the proton's vortex. For example:

A. Collisions with protons:

#1 if the positive fourth dimensional vortex tries to break off to form a neutral particle and another positive particle, it does not have the correct quark content. When the up, up, down of the proton is added to the up, anti-down of the tunneling positive pion there is no way to create just a neutral and positive particle from the five quarks. [Note, although at first glance it appears as if the up, up, up, could form a delta-two resonance particle while the down, anti-down could create the neutral pion, this cannot happen. It cannot happen because the tunneling positive pion has no fourth dimensional vortex to add to the proton's 4d vortex to create the plus two charge of the delta-two resonance. Hence, the delta-two resonance cannot be created.] Consequently, it appears as if no resultant particles are created as a consequence of a collision with a proton.

However, if the tunneling positive pion collides with an anti-proton with the proper momentum it will create resultant particles. Because the anti-proton consists of an anti-up, anti-up, anti-down set of quarks, when added to the tunneling positive pions up and anti-down quarks, we suddenly have the right combination. The anti-down, anti-down, and anti-up quarks create the anti-neutron; the anti-up creates the negative muon; and the up quark creates the muon neutrino. This reaction conserves charge [remember the word *positive* in the tunneling positive pion has no 4d vortex, hence, the tunneling pion has no charge; the word *positive* is only a designation used to distinguish it from the *negatively charged quarks in the* tunneling pion]; conserves the conservation of baryons; and the conservation of lepton number.

B. Collisions with neutrons:

If the tunneling positive pion strikes the neutron, the breakup of the neutron loop limits the resultant particles that can be created. For example:

#1 if the tunneling positive pion strikes the neutron the quark combination will consist of an up, down, down, up, anti-down. Because the neutron vortex is caught in a loop, it can only break exposing its negative and positive ends or break again creating another loop limiting the possible particles that can be created. Since no combination of the five quarks creates a positive and negative particle (one of which has to be a baryon), the tunneling positive pion cannot create particles via striking a low energy neutron.

However, if the tunneling positive pion collides with an anti-neutron of the proper momentum, suddenly, the right combination is present. The three anti-downs create an anti-omega minus particle (+1), while the anti-up creates the negative muon (-1), and the up tunnels, creating the muon anti-neutrino. This reaction conserves charge, baryon number, and lepton number.

C. Collisions with positrons:

If the tunneling positive pion strikes a low energy positron it can create a positive pion.

Chapter 39
Schematics of Higher Dimensional Particles

> The relationship between the 3d holes, the 4d vortex, and the 4d holes and their 5d vortex can be drawn schematically. Below are schematic drawings of positive and negative muons.
>
> Note: for non-scientists, according to 20th Century science, a Muon appears to be nothing more than a heavy electron. But this is not true! A muon possesses a quark!

Figure 39.1

POSITIVE MUON   NEGATIVE MUON

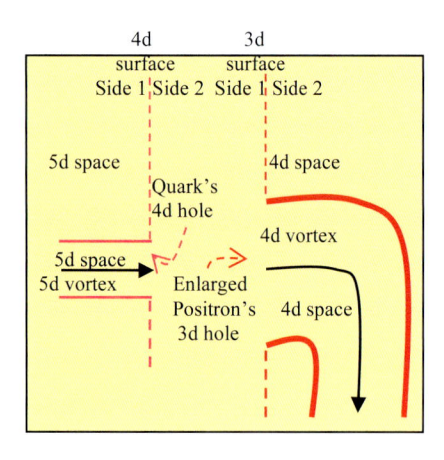

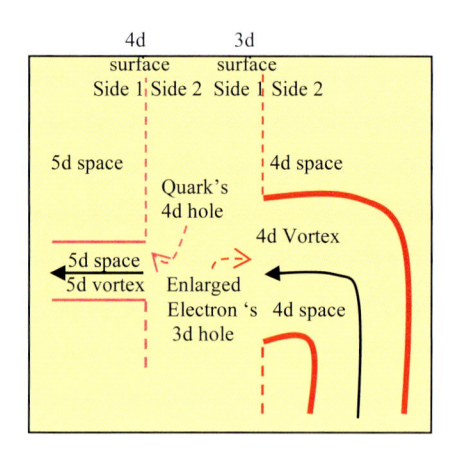

In Figure 39.1, the thick red dashed line represents the 3d volume of space we live in, while the dashed purple line represents the 4d volume of space. Note how both the electron's 3d hole and the positron's 3d hole have enlarged because of the presence of the quark. Note too, how the surface of the 4d vortex is constructed out of 3d space surrounding the volume of 4d space within its interior; and, that the surface of the 5d vortex is constructed out of 4d space surrounding a volume of 5d space within its interior. The black arrows represent the direction space is flowing in the vortices.

It is important to remember too, that the 4d hole exists within the 4d space found within the 3d hole. However, here, it is only drawn apart from the 3d hole so that it can be more clearly seen and to make the schematic less confusing. [Later on, this schematic drawing will be expanded to include all six quarks.]

It should also be mentioned again as originally explained in *The Vortex Theory* that as the hole in space expands in size, the space surrounding it becomes less dense decreasing the volume of the flow. This decrease in density creates an inverse relationship between the size of a hole and the volume of space flowing into it: causing larger holes to possess the same volume of flowing space as smaller holes – causing different sized holes to all possess the same charge that the electron or positron originally had.

Chapter 40
How Particle Collisions Create New Particles

> One of the most fascinating areas of particle physics are particle collisions and decays. It is incredible how single particles can collide and decay into many new and different particles that seem to have almost nothing in common with the original particles that spanned them. However, before we can explain exactly how these most curious and wonderful decays of quarks occur we must first explain how quarks are created.

When two particles such as an electron and a positron are accelerated to higher and higher velocities in a particle accelerator, their momentum is increased. But what is **momentum?**

Classical physics defines momentum as a formula [P = mV]; where a mass [m] is multiplied by its velocity [V]. But what creates its velocity? How does a "particle" move when it is really a hole in space?

The only way a hole can move through space is for the space in front of it to split apart, move around the hole, then come back together at the back of the hole: creating the effect of a quantized wave. This wave is different from the theorized aether waves proposed a hundred years ago. A quantized wave is created upon the interior surface of the hole, while an aether wave is similar to the wave created by the bow of a boat: where the surrounding space is pushed to either side.

When an object in motion constructed out of particles made of holes strikes another object at rest, the collision imparts some of the quantized waves to the interior surfaces of the holes in the struck object. These waves cause the holes in the object at rest to move; the loss of some of the waves in the moving object causes it to slow down. Or a better way to explain it is by using the term "amplitude."

Some of [or all of] the amplitude of the waves in the holes of the moving object, are transferred to the holes in the stationary object: "accelerating" it. This transference explains the **conservation of momentum**.

This increase in momentum creates a distortion of the spherical shape of the particle into a pear shape.

Figure 40.1

The shape of the electron or positron at rest:

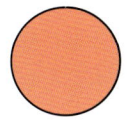

Note: this is an ideal shape. The shape of any spherical hole is never exactly a sphere because there is always a gravitational force somewhere that is distorting its surface.

Figure 40.2

The shape of the electron when moving at a high velocity **V**:

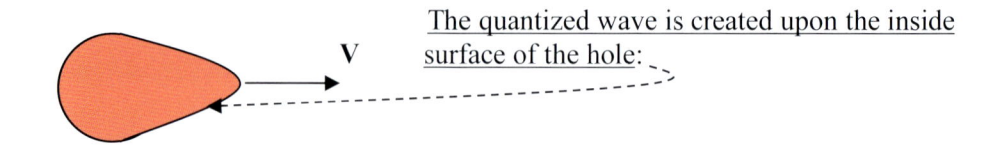

The quantized wave is created upon the inside surface of the hole:

Note too: this distortion is greatly exaggerated. Although it might be possible to greatly distort the shape of the electron, positron, or any other charged particle as much as is shown here, the actual distortion of the hole is much less.

When an electron and positron collide at slow velocities, they annihilate each other creating two gamma rays: [the two volumes of 3d space that existed within their interiors is expelled]. And just as the two dimensional surface of the water is thrown upward into the three dimensional volume of the air when two waves on the ocean collide, when these particles collide, their combined 3d distortions are suddenly and violently transferred into the 4d volume distorting it too. This opens up additional holes within the 4d volume of space that is also the surface of 5d space. These holes are up and down quarks.

If the momentum is not high the distortions are transferred only into side 2 creating only the up, anti-up quark pair: creating a muon anti-muon pair. If the momentum is greater, the distortion is transferred into both side 1 and side 2 simultaneously, creating the down, anti-down quark pair: creating Pions. Depending upon the momentum of the colliding particles, even higher dimensions of space are penetrated. When higher dimensions of space are penetrated, the strange and anti-strange pairs are created. Then the charm and anti-charm pairs are created; next come bottom and anti-bottom pairs; until finally the top and anti-top pairs are created. Of course, these quarks then combine to form the many different types of leptons, mesons, and baryons.

A similar situation occurs when a fast moving particle strikes the nucleus of an atom. Again, the lower dimensional surfaces of higher dimensions of space are distorted creating holes in higher dimensional space. These higher dimensional holes become quarks creating new particles. The same is true for higher energy photons striking the nuclei of atoms.

When high energy photons such as gamma rays strike the nuclei of atoms, the very dense and fast vibrating space of the gamma ray violently distorts the surface of 3d space, that in turn distorts the 4d volume, creating holes within the fourth dimensional volume. These holes then combine to form new particles.

The above knowledge now sets the stage for one of the most fascinating phenomenon in the world of subatomic particle physics – the decay of particles and how quarks change "Flavor".

THE DIFFERENCE BETWEEN STRONG FORCE AND WEAK FORCE CREATIONS

The difference between the "strong force" creating a set of particles and the "weak force" is found in the method of distorting space. The strong force particle creations occurs when two surfaces collide or annihilate during particle anti-particle decays; while weak force creations occur when the space flowing within a vortex pushes out into the other side of space.

The "strong force" creates both the particle hole and anti-particle hole at the same time. This effect is created by an indentation in space with entrance and exit holes which are roughly a half donut shape Figure 40.3 [#1]. As this indentation increases in size, the elasticity of the space tries to make the holes close back together again Figure 40.3 [#2]. And as they close, the two ends of the vortex are created, and the inward and outward flow of the surface causes the 4d space to begin to flow in and out of the holes creating the two "particles" Figure 40.3 [#3].

This bend is called a "Mutual Creation" because it creates a particle anti-particle combination.

Figure 40.3   "Mutual creation":

#1

#2

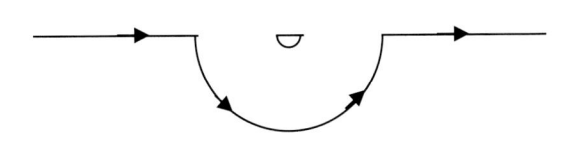

#3

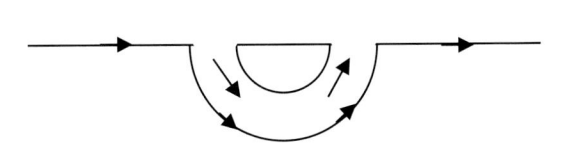

The "weak force" tends to create a bend of space in a different manner. The weak force tends to be created by a *volume* of space from one side of space pushing or pulling into the opposite side of space: side 1 pushing into side 2; side 2 pushing into side 1; side 1 pulling into side 2; or side 2 pulling into side 1.

Because it is not strong enough to create a big indentation creating both particles at once, it pushes into the opposite side of space creating a single hole [#1]. The opposite end pops back out of space becoming a neutrino {#3}. Note: the neutrino is traveling at C, the speed of light the same as the arrows in the vortex.

Although the above drawings are merely schematic drawings, they create the stage for the presentation of a most fascinating topic: Tunneling Particles.

Chapter 41
The Explanation of How Quarks Change "Flavor"

> Perhaps one of the most fascinating though perplexing questions in all of particle physics is how quarks can change from one type of quark into another type of quark. Perhaps too, early quark theorists whimsically designated this process "changing flavor" because they had absolutely no idea what was causing it. For example, what mechanism in nature can change a quark with a +2/3 charge change into a quark with a lesser -1/3 charge; or, and equally perplexing, change a quark with a -1/3 charge into a completely different quark possessing a much greater +2/3 charge?

The search for an explanation for this change in flavor must have been very frustrating for those early quark pioneers of the 1960's. They believed they knew what was happening, but they had absolutely no idea how it was happening. They didn't even know how to tell other physicists about it.

How do you explain a phenomenon that has no similar phenomenon in nature and even worse, seems to defy logic itself! For if quarks were compared to apples, oranges, peaches and pears, you are dealing with a situation where a pear could turn into a peach, and a peach that could turn into an orange, and an orange that could turn into an apple! And today's physicists are no better off.

Fifty years later, nothing seems to have changed. Today's physicists are equally at a loss to explain to their students how such a most unusual and bizarre situation can occur.

Fortunately, with the discovery of the Vortex Theory, the explanation is now quite simple. The key to understanding how a quark with one type of charge can decay into another type of quark with an entirely different charge is easily understood by using the concept of the two sides of space and their respective elasticity's.

WHY QUARKS DECAY

Quarks decay because they become unstable holes in space.

The collapse of a quark begins when the elasticity of the space it is created within is exceeded. When this situation occurs, the space the hole is created within creates a pressure that contracts and closes the hole back up. This procedure is analogous to throwing a heavy bowling ball into a swimming pool. When the bowling ball hits the water, it suddenly pushes the water aside creating a momentary hole as it continues on, penetrating into the volume of water. This hole only exists for an instant because as soon as the bowling ball penetrates the surface, the water tries to instantly rush back in; filling up the hole, closing it back in behind the heavy ball. This is very similar to how quarks are created and collapse.

In the previous chapter, we saw how sudden distortions in the surface of space created distortions within the volume of space and created holes – or quarks – within this volume. Since each volume of space is the surface of the next higher dimension of space, these new holes [quarks] then combined to form new particles. However, many of these new particles only last for very short periods of time because the quarks within them collapse [decay].

The reason why these quarks collapsed or rather decayed is due to the elasticity of space. Quarks are originally created when bent regions of lower dimensional space suddenly collide. The collision twists space so violently that its elasticity is exceeded. This creates a rip or tear in the volume of the next higher dimensional space. The tear spreads space apart creating a hole in it. However, once the higher dimensional hole is created, and the force of the lower dimensional bent space has dissipated, the elasticity of the higher dimensional space takes over. The elasticity of the higher dimensional space now pushes [or pulls] backwards against the hole.

If there is nothing to counteract this sudden push or pull of the elasticity of this particular dimension of space, the hole collapses instantly [at the speed of light]. This instant collapse of the hole creates a push or pull that is instantly transferred to the opposite side of space.

This instant push or pull, in turn, pushes or pulls upon the opposite side of space opening a new set of holes in its surface. If the hole that collapsed was on side 1 with a charge of -1/3, the new hole opened directly opposite to where it was and has a charge of +2/3. Or, if the hole that collapsed was on side 2 with a +2/3 charge, the new hole that opened directly opposite to where the collapsed hole was will now have a charge of -1/3. But this is just the beginning.

Because individual quarks are only the ends of vortices, when one end is suddenly created, another end is always formed with it creating another quark or particle. Furthermore, because quarks exist within quarks, when a higher dimensional quark decays, the one it resided within is usually left – causing three seemingly new particles to be miraculously created. To better understand how this process works, some of the possible particle decays are presented.

Each of the particle decays represents one type of decay mode. At the time of this writing there appear to be only these type of decays. However, as more work is devoted to this area of research perhaps more decay modes will be discovered. The liberty was taken to give names to these decays, so it is easier to refer to each individual type later.

TYPES OF DECAY

In the next drawings, a particular quark is drawn on the left side of the page, and one particular decay mode is drawn on the right.

Figure 41.1 "CLASSICAL DECAY"

The "Classical Decay" of the charm quark is examined below. In the classical decay of the charm quark, three quarks are created: two are joined by a vortex of flowing space on the opposite side of the decay; and one is on the same side of the decay.

In the below drawing notice how the Charm quark is located on the 5d surface of the 6d volume of space. Notice too how it resides within an enlarged Up quark. It is also important to realize that the Charm quark exists upon side 2 of space. [The red line represents the same surface in each drawing.]

Charm quark decays into the Down quark; Anti-down quark; Up quark

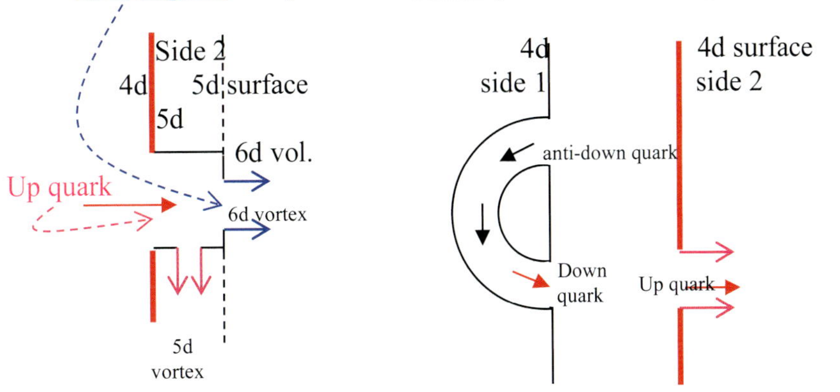

When the charm quark decays, it collapses *pulling* side 1 outward creating the down quark and the anti-down quark. After its collapse, the up quark that it resided in is left by itself on side 2.

To be able to predict if the down or anti-down quark is created first, observe the red arrows. Note how the red arrow that represents the direction of the flow of space into the charm quark would pull side 1 outward during the charm quarks collapse, creating the down quark first (4d space is flowing out of the down quark}. This important observation will become more meaningful to us later on.

Also note the red arrow in the up quark (4d space flows into the up quark). This red arrow indicates the direction of the flow of space into side 2. The direction of this arrow is important because it reveals that the quark left over after the charm collapsed was the up quark and not the anti-up quark (4d space flows out of the anti-up quark). The direction of this arrow is found by looking at the direction of the red arrow in the charm quark.

It must also be mentioned that the 6d blue vortex of the charm quark "deflates". It deflates backward in the direction of the anti-charm quark that exists within another particle and is not shown in this drawing.

Notice too how the purple arrows of the up quark's 5d vortex in the charm quark can now be seen in the right hand drawing. Even though the 6d vortex of the charm quark is deflating, the 5d vortex of the up quark survives the charm quark's destruction.

Figure 41.2 "CLASSICAL DECAY"

The "Classical Decay" of the strange quark is examined below. In the classical decay of the strange quark, again, three quarks are created: two are joined by a vortex of flowing space on the opposite side of the decay; and one is on the same side of the decay.

In the below drawing notice how the strange quark is located on the 5d surface of the 6d volume of space. Notice too how it resides within an enlarged down quark. It is also important to realize that the strange quark exists upon side 1 of space. [The green line represents the same surface in each drawing.]

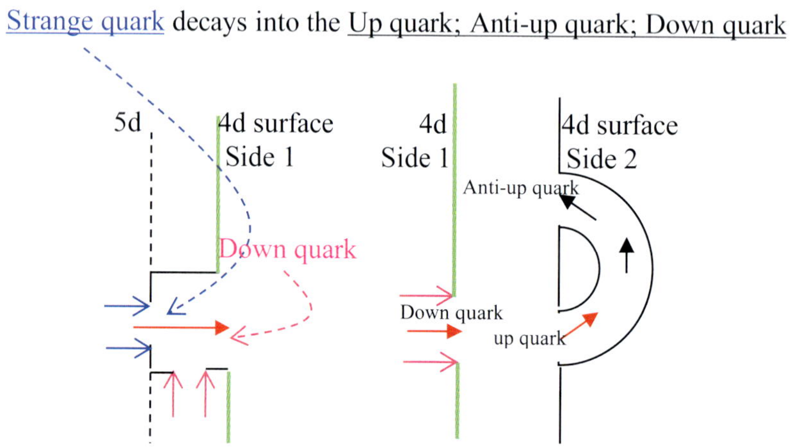

Strange quark decays into the Up quark; Anti-up quark; Down quark

129

When the strange quark decays, it collapses *pushing* side 1 outward creating the up quark and the anti-up quark. After its collapse, the down quark that it resided in is also left.

To be able to predict if the up or anti-up quark is created first, observe the red arrows. Note how the red arrow that represents the direction of the flow of space out the strange quark would push side 2 outward during the strange quark's collapse. This arrow indicates that the up quark is created first and not the anti-up quark. [This important observation will also become more meaningful to us later on.] Also note the direction of the red arrow in the down quark. This red arrow is found by observing the direction of the red arrow in the strange quark. This red arrow is important because it indicates that space is flowing out of the quark the charm resided in, identifying it as a down quark. [Space is flowing out of the down quark.]

The 6d blue vortex of the strange quark ceases to exist in the second drawing. It is gone because when the strange quark "deflates," its vortex deflates backward in the direction of the anti-strange quark that exists within another particle and is not shown in this drawing.

Notice too how the purple arrows of the down quark's 5d vortex in the strange quark can now be seen in the right hand drawing. Even though the 6d vortex of the strange quark is deflating, the 5d vortex of the down quark survives the strange quark's destruction.

Figure 41.3 "TOTAL DEFLATION"

During a "Total Deflation," the decaying quark expels all of its energy directly into the opposite side of space. Two new quarks are created that are joined together by a vortex. In the example below, a down quark decays into an up quark and an anti-up quark.

The down quark decays into an up and anti-up quark.

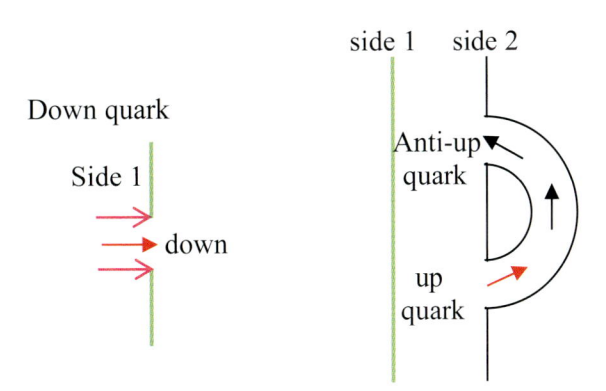

Because space is flowing out of the down quark [red arrow], when it decays, it pushes outward into side 2 creating the up quark first [red arrow].

When the down quark's vortex deflates, it collapses backward through the volume of 5d space in the direction of the anti-down quark it is connected to within another particle. Therefore, the purple vortex is not seen on side 2 (the green line) in the second drawing.

Figure 41.4 "BREAKAWAY DECAY WITH NEUTRINO"

In the breakaway decay with neutrino, the anti-down quark deflates creating the muon neutrino.

When the anti-down "deflates," the vortex **tunnels backward** in the direction of the down quark it is connected to in another particle. When the vortex collapses, 4d space is pulled outward in the direction of side 1 creating an indentation that becomes the muon neutrino.

In the below drawings, the anti-down quark exists within a positive pion made up of an up quark and an anti-down quark.

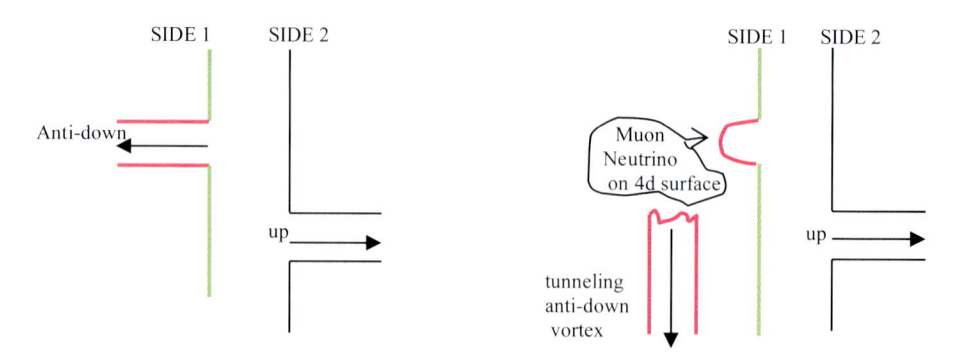

Figure 41.5 AN IMPORTANT OBSERVATION REGARDING NEUTRINO CREATION

When quarks decay, they can bend the opposite side of space inward or outward creating neutrinos. The direction of these inward or outward bends in space is determined by the direction space is flowing in the quarks. For example, when down, anti-down, and up, anti-up quarks decay, side 2 is pushed both inward and outward creating both a neutrino and an anti-neutrino:

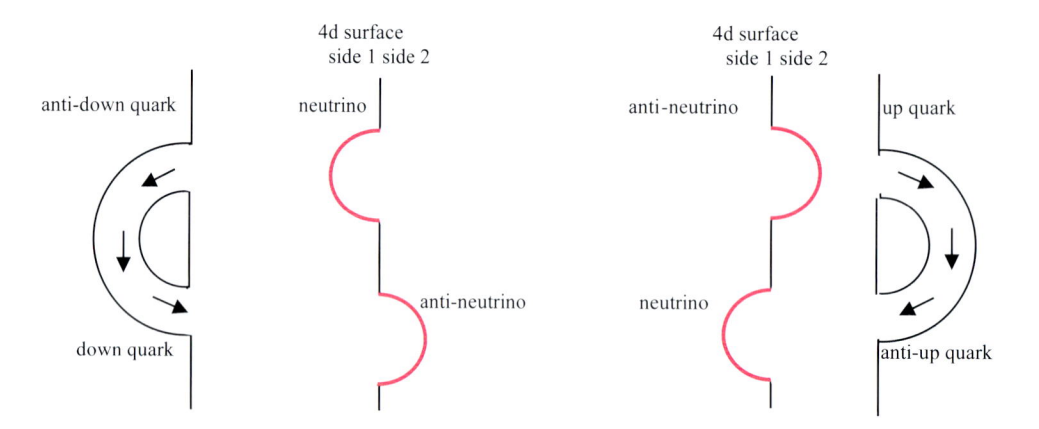

In the above figures, notice how the arrow in the anti-down quark points to side 1, and also, how the neutrino bulges outward towards side 1. If side 1 was represented by a positive sign (+), both the anti-down quark and the neutrino could be represented by a positive sign; if side 2 was represented by a negative sign (–), then both the down quark and the anti-neutrino could be represented by a negative sign.

Next, notice how the arrow in the anti-up quark points to side 1, and also, how the neutrino bulges outward towards side 1. If side 1 was represented by a positive sign (+), both the anti-up quark and the neutrino could be represented by a positive sign; if side 2 was represented by a negative sign (–), then both the up quark and the anti-neutrino could be represented by a negative sign. [These observations are about to become very significant].

Chapter 42
Lepton Creation During the Decay of Positive and Negative Pions

> The reason why the muon neutrino is created during the decay of the positive pion, and the anti-muon neutrino is created during the decay of the negative pion will be examined in this chapter. These relationships are important because they reveal how the space flowing in a vortex pushes or pulls against the side of space opposite to it during its decay. And it is this pull or push that bends the opposite side of space inward or outward creating the muon neutrino or muon anti-neutrino.

The relationship between the direction of the space flowing into or out of a vortex, and the bends they create in the opposite side of space are responsible for the creation of rules #3 and #4 in Table 52-1, Chapter 52: two of the eight rules responsible for the creation of the law of the conservation of lepton number.

THE DECAY OF THE POSITIVE PION

Figure 42.1 The deflation of the anti-down quark creates a muon neutrino: V_u

In the below drawings, the decay of the anti-down is shown during the decay of a positive pion π^+. Note too: the positive pion is constructed out of an Up quark and an Anti-down quark.

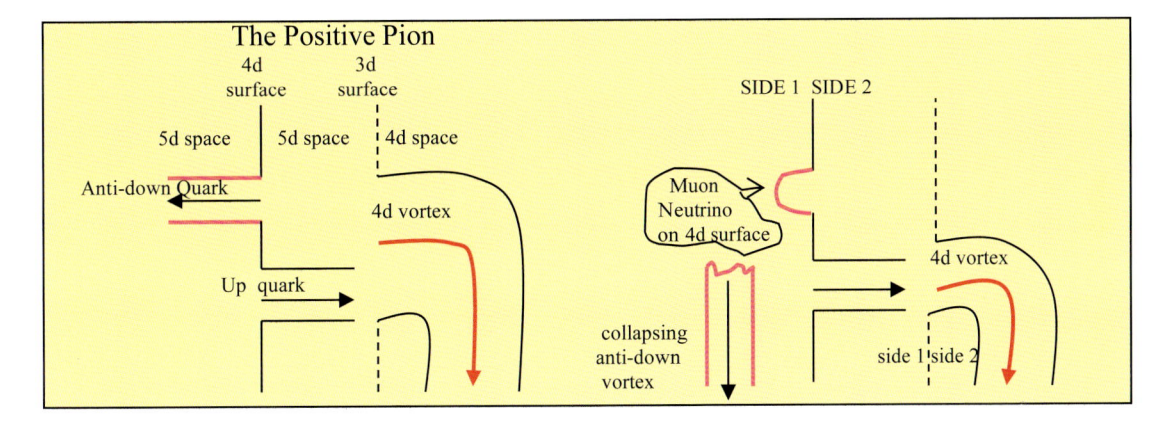

When the anti-down quark deflates, its inward pull [note the direction of space (towards side 1) flowing in the anti-down quark] into side 1 pulls side 2 outward. This inward pull creates a quantized volume of 5d space bent into side 1 that becomes the Muon Neutrino. The anti-down vortex collapses backward towards the down quark (not shown in drawing) that is at the other end of this vortex. The removal of the 5d volume of space previously occupied by the decayed anti-down quark causes its 4d surface of the 5d volume [the 4d volume within the pion] to be reduced in size, forcing the 4d vortex to reduce in size creating the positive muon. This reaction is represented as the **equation 1**: $\pi^+ \rightarrow V_u + \mu^+$

CREATION OF CONSERVATION OF LEPTON NUMBER:

In Figure 42.1, note that the deflating anti-down quark pulled the surface of 4d space inwards, towards side 1, in an opposite direction to that of the space flowing in the 4d vortex [red arrow towards side 2] of the positive pion. This bend in 4d space *towards side 1*, [previously designated +1], gives the muon neutrino a lepton # of +1; and the direction of the flowing space in the positive muon vortex *towards side 2*, [previously designated -1], gives the positive muon a lepton # of -1. Because the positive pion has a lepton # of 0, when all three numbers are plugged into the equation 1, **0 = +1 -1**, we see that the conservation of lepton number is conserved. But even more important, for the first time we suddenly begin to see how the conservation of lepton number is being generated!

We have also just seen one of the reasons why a muon neutrino is created along with the creation of a positive muon [see Chapter 52, Table 52-1, #3]. Next, we shall see how an anti-muon is created during the creation of a negative muon.

THE DECAY OF THE NEGATIVE PION

Figure 42.2 In a negative Pion, the decay of the down quark creates an anti-muon neutrino $\overline{V}_u$.

In the below drawings, the decay of the anti-down is shown during the decay of the negative pion π-. Note too: the negative pion is constructed out of a down quark and an anti-up quark.

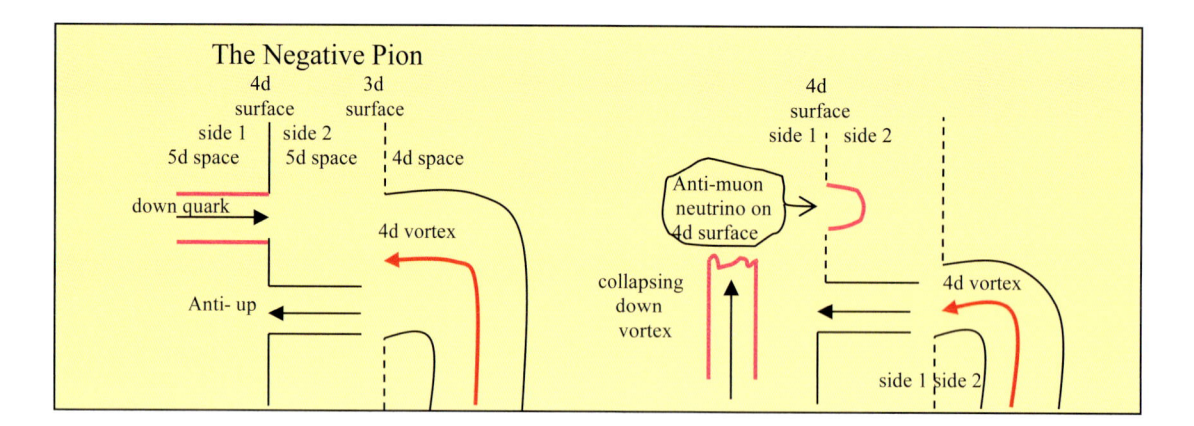

When the down quark deflates, its outward flowing 5d space into side 2 [note the direction of the arrow in the down quark] pushes side 2 inward. This inward push creates a quantized volume of 5d space bent into side 2 that becomes the muon anti-neutrino. [Note how the muon anti-neutrino is created upon the surface of 4d space.] The removal of the 5d volume of space causes its 4d surface [the 4d volume within the pion] to reduce in size, forcing the 4d vortex to reduce in size becoming the negative muon.

This reaction is represented by **equation 1**: $\pi^- \rightarrow \overline{V}_u + \mu^-$

CREATION OF CONSERVATION OF LEPTON NUMBER:

Note that the decaying down quark pushed the surface of 4d space outward towards side 2 in an opposite direction to that of the space flowing in the 4d vortex of the negative pion [red arrow towards side 1]. This bend in 4d space *towards side 2* [previously designated -1] gives the muon anti-neutrino a lepton # of -1; and the direction of the flowing space in the negative muon vortex *towards side 1* [previously designated +1] gives the positive muon a lepton # of +1. Because the positive pion has a lepton # of 0, when all three are plugged into the above equation 1: **0 = -1+1** hence, the lepton number is conserved.

We have also seen the first reason why a muon anti-neutrino is created during the creation of a negative muon . Next, we shall see how a muon anti-neutrino, and an electron neutrino are created during the decay of a positive muon.

Chapter 43
Neutrino Creation During the Decay
of the Positive Muon

LEPTON RELATIONSHIPS

> During the decay of the positive muon, the reason why the muon anti-neutrino and the electron neutrino are both created will be examined. These relationships are important because they reveal how the space flowing into a decaying vortex pulls the opposite side of space outward creating an anti-neutrino. It also reveals how a less dense region of space pulled towards a flowing vortex suddenly allows side 2 to push outward into side 1, creating an electron neutrino.

These relationships are examples responsible for the creation of rules #7 and #1 in Table 52-1 in Chapter 52: two more of the eight laws responsible for the creation of the conservation of lepton number.

Figure 43.1 In the positive muon, the deflation of the up quark creates a muon anti-neutrino: $\overline{V_u}$

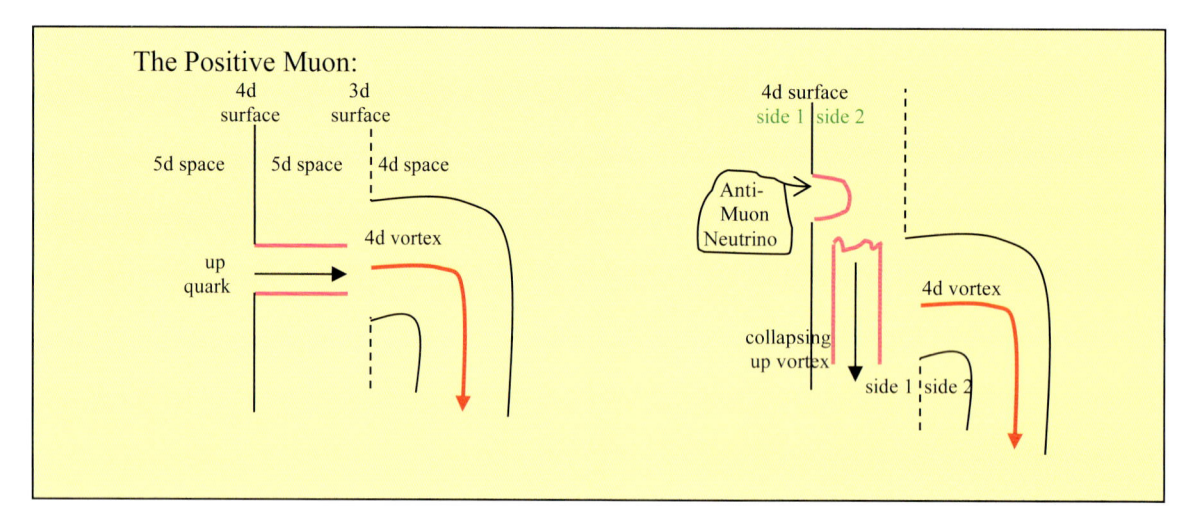

The decay of the positive muon takes place in two parts. The first part of the decay occurs when the up quark deflates, and its inward flow - into side 2 - pulls side 2 inward. This inward pull creates a volume of space bent into side 2 that becomes the muon anti-neutrino.

The second half of the decay is created by the deflation of the extra volume of less dense space in the volume of the 4d vortex needed to decrease the density of the space in the vortex. [as seen in figure below] This extra volume of 4d space within the positive muon was needed to change the charge of the up quark from a value of +2/3 to a value of +1. However, unlike the decays of the up, down, and anti-down quarks that all bend space in the direction of the space flowing in their vortices; the decay of this extra volume of the 4d space bends the 3d surface of space outward in the *opposite direction* to the flow of the space in the vortex.

The reason for this reversal comes from the side 1 the less denser region of three dimensional space surrounding this hole. When space is pulled into the hole, a less dense region of side 1 space surrounds the three dimensional hole. The same is true for the interior. Inside the three dimensional hole, the fourth dimensional space surrounding the fourth dimensional hole is pulled inward into the fourth dimensional hole of the up quark also creating a less dense region of space.

When the up quark decays, the less dense region springs backward in the opposite direction pushing outward towards side 1 of space. As it does, the now denser side 2 immediately pushes into this less dense region of side 1. This outward distortion creates a volume of space bent into side 1 that becomes the electron neutrino; the shrunken vortex becomes the positron:

Figure 43.2

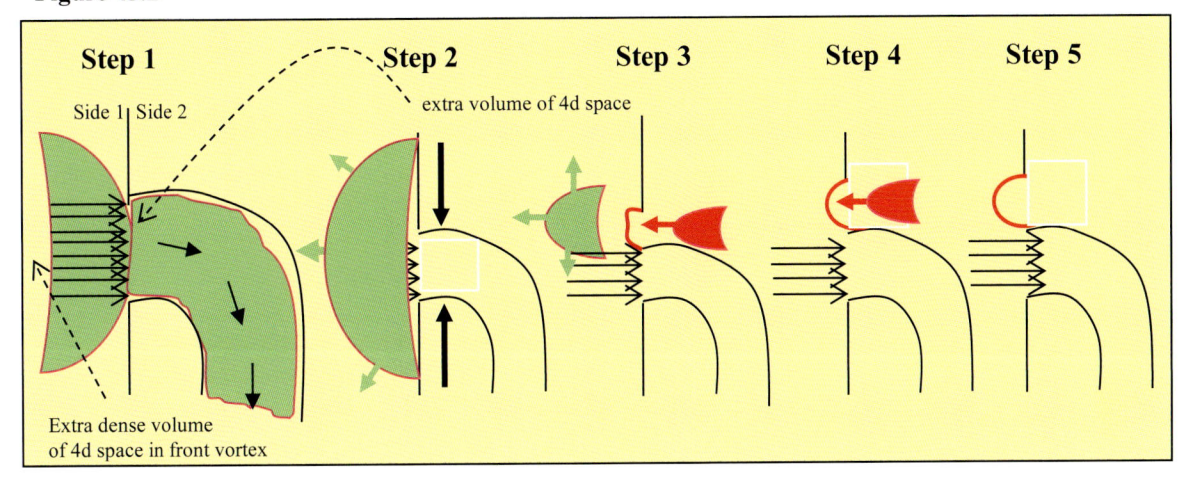

As seen in the above figure, the extra volume of *dense* 4d space being pulled into the vortex [crescent shaped shaded green area in step 1] of the positive muon suddenly reverses direction when the vortex collapses [step 2]. It now pushes back outward into the 4d volume of side 1 [step 3 & 4] allowing side 2 to push into the indentation; this sudden outward shove from side 2 towards side 1 pushes a region of side 1 outward in the opposite direction [step 4]. This outward push creates a volume of side 2 space bent into side 1 that becomes the electron neutrino V_e [step 5] and the shrunken vortex becomes the positron.

It is important to note that the muon neutrino possesses more mass than the electron neutrino because the muon neutrino possesses a tunneling up quark while the electron neutrino is merely a surface distortion.

Figure 43.3 The simplified version of the creation of the electron neutrino: V_e.

To simplify the sequence of events pictured in Figure 43.2 & Figure 43.3 is now used. The green volume of space in front of the vortex represents the less dense volume of space that reverses direction and pushes backward into the 4d volume of side 1 pulling side 2 outward, creating the electron neutrino.

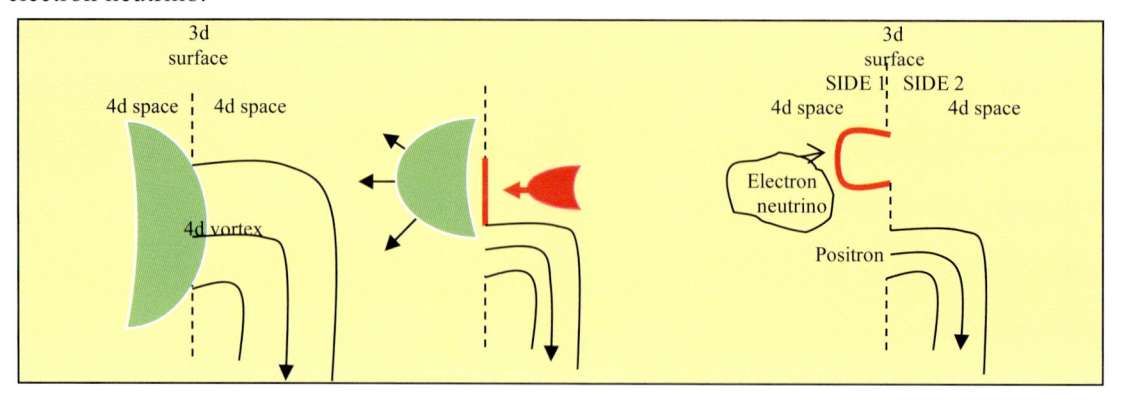

This reaction is represented as the **equation:** $\mu^+ \rightarrow \overline{V_u} + V_e + e^+$

CREATION OF CONSERVATION OF LEPTON NUMBER:

In the Figure 43.1, note how the direction of the space flowing in the decaying up quark's vortex pulled the surface of 4d space inward towards side 2. Note too that this direction is the *same* as the direction of the space flowing in the 4d vortex of the positive muon [towards side 2]. This bend in 4d space towards side 2 [previously designated -1] gives the muon anti-neutrino a lepton # of -1; and the direction of the flowing space in the positive muon vortex towards side 2 [previously designated -1] gives the positive muon a lepton # of -1.

The creation of an anti-muon neutrino with a positive muon is an example of rule #7 Table 52-1.

Next, in Figure 43.3 [the second half of the decay], notice how the extra volume of space located just outside of the positive muon, [the volume needed to change the charge of the up quark from +2/3 to a value of +1], pushes backward into the 4d space from which it was pulled. This backward push pulls the 3d surface inward creating the electron neutrino. Observe too, how this outward push is in the opposite direction to the space flowing in the positive muon's vortex. Because this same vortex sheathing the positive muon deflates to become the positron's vortex, the direction of the flow of the positron's vortex is oriented in a direction opposite to the direction that the electron neutrino is oriented. This decay gives us our first insight into why an electron neutrino is paired with a positron when both are on the same side of the equation [affirming rule #1 in Table 52-1].

[Note too, in contrast to the muon anti- neutrino that is formed upon the 4d surface and bent into the 5d volume of space, the electron neutrino was created upon the surface of 3d space and is bent into the 4d volume.]

And finally, without looking at our table of lepton numbers, but instead, just by looking at the direction the neutrinos are bent into the two sides of space and the direction the space is flowing within the lepton vortices, we come up with the following lepton numbers: positive muon = -1; muon anti-neutrino = -1; electron neutrino = +1; and the positron = -1. When all four are substituted into the equation representing this decay, $\mu^+ \rightarrow \overline{V_u} + V_e + e^+$ the results are: -1 = -1 +1 -1 And we find that we have come up with the exact same result we would have come up with had we used our table. Next, we shall see how an anti-muon is created during the decay of a negative muon.

Chapter 44
Neutrino Creation During the Decay of the Negative Muon

LEPTON RELATIONSHIPS

> During the decay of the negative muon, the creation of both the muon neutrino and the electron anti-neutrino are examined. The decays will also give examples of how rules #8 and #2 of Table 52-1 are created.

Figure 44.1 The decay of the anti-up quark creates a muon neutrino:

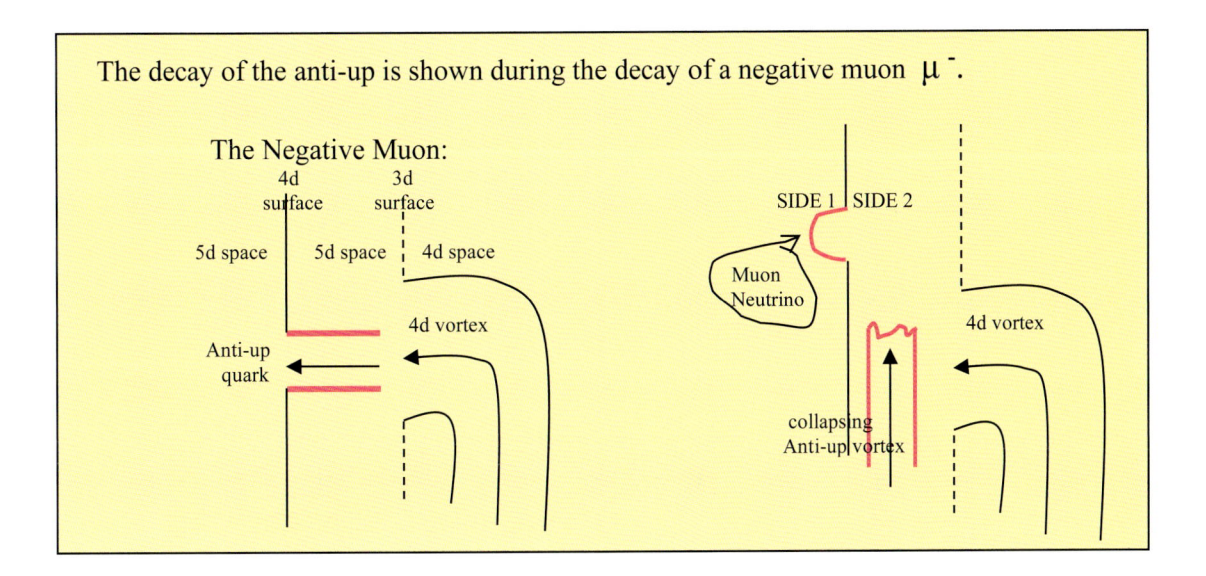

When the anti-up quark deflates, its outward flow – towards side 1 – pushes side 1 outward. This outward push creates a volume of space bent into side 1 that becomes the muon neutrino. However, this is only the first half of the decay.

The second half of the decay is created by the deflation of the extra volume of space in the volume of the 4d vortex needed to increase the density of the space in the vortex [see Figure 44.2 below]. This extra volume of 4d space within the negative muon was needed to change the charge of the anti-up quark from a value of -2/3 to a value of -1]. However, unlike the decays of the up, down, and anti-down quarks that all bend space in the direction of the space flowing in their vortices; the decay of this extra volume of the 4d space bends the 3d surface of space outward in the *opposite direction* to the flow of the space in the vortex.

This reversal occurs because unlike the collapse of the quark vortices, the 4d vortex does not collapse completely. [If it had collapsed completely, the resultant bend in the 3d surface would be a single combination of these two opposite bends.] Consequently, the extra volume of 4d space flowing out of the negative muon suddenly reverses its flow. As the size of the vortex collapses, this sudden reversal pushes outward upon the 3d surface distorting it outwards towards side 2. This outward distortion creates a volume of side 1 space bent into side 2 that becomes the electron anti-neutrino; the shrunken vortex becomes the electron.

Figure 44.2

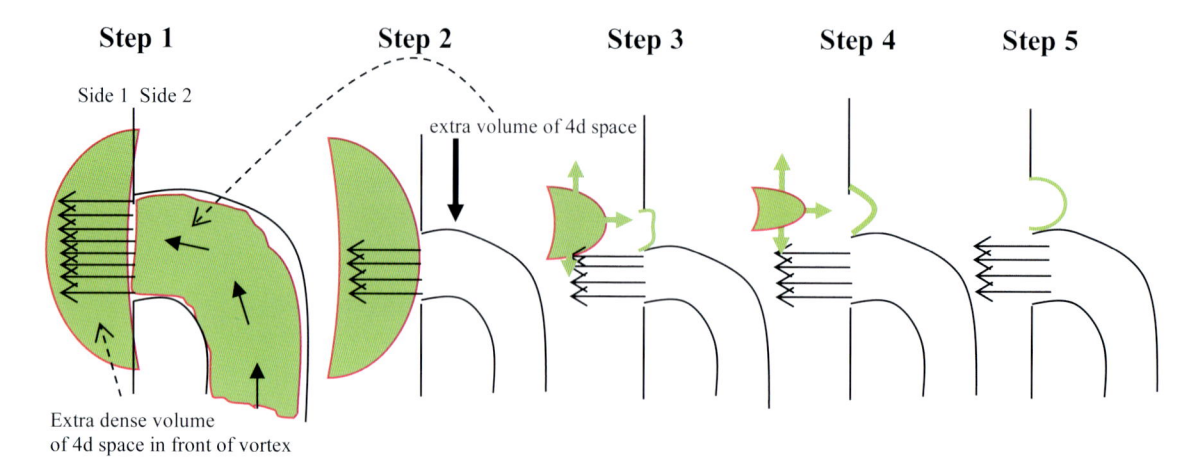

As seen in Figure 44.2, the extra volume of dense 4d space being pushed out of the vortex [crescent shaped shaded area in step 1] of the negative muon reverses direction [step 2] when the vortex collapses. It suddenly pushes back outward into the 4d volume of side 2 [step 3]; this sudden outward push from side 1 towards side 2 pushes side 2 outward in the opposite direction. And again, this outward push into side 2 creates a volume of space bent into side 2 that becomes the electron anti-neutrino, and the shrunken vortex becomes the electron.

Figure 44.3 Creation of the Electron Anti-neutrino $\overline{V}_e$:

The creation of the electron anti-neutrino can again be seen in the drawing below:

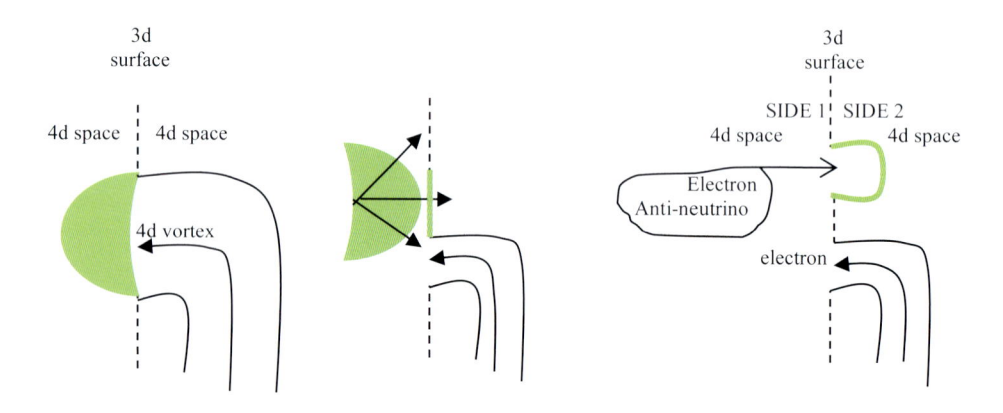

This reaction is represented as the **equation**: $\mu^- \rightarrow V_u + \overline{V}_e + e^-$

CREATION OF CONSERVATION OF LEPTON NUMBER:

In Figure 44.1, note how the direction of the space flowing in the decaying anti-up quark's vortex pushed the surface of 4d space outward towards side 1. Note too that this direction is the *same* as the direction of the space flowing in the 4d vortex of the negative muon [towards side 1]. This bend in 4d space towards side 1 [previously designated +1] gives the muon neutrino a lepton # of +1; and the direction of the flowing space in the negative muon vortex towards side 1 [previously designated +1] gives the negative muon a lepton # of +1. This decay also reaffirms Rule #8 in Table 52-1.

Next, in Figure 44.3, [the second half of the decay], notice how the extra volume of space pushing outward into the 4d volume from the negative muon, [that was needed to change the charge of the up quark to a value of +1], reverses direction and pushes outward into side 2 creating the electron anti-neutrino. Observe how the anti-neutrino's outward bulge is in the opposite direction to the space flowing in the negative muon's vortex. Because this same vortex sheathing the negative muon deflates to become the electron's vortex, the direction of the flow in the electron's vortex is oriented in a direction opposite to the direction that the electron anti-neutrino is oriented. And again, not only can we understand, we can also *see* why an electron anti-neutrino is paired with an electron when both are on the same side of the equation. This also gives an explanation for rule #2 in Table 52-1.

[Note too, in contrast to the muon neutrino that is formed upon the 4d surface and bent into the 5d volume of space, the electron anti-neutrino was created upon the surface of 3d space and is bent into the 4d volume.]

And finally, without looking at our table of lepton numbers, but instead, just by looking at the way the leptons are bent into the two sides of space we come up with the following lepton numbers: negative muon = +1; muon neutrino = +1; electron anti-neutrino = -1; and the electron = +1. When all four are substituted into the equation representing this decay,

$\mu^- \rightarrow \overline{\nu}_\mu + \nu_e + e^-$ and the results are: +1 = +1 -1 +1. And we find that we have come up with the exact same result we would have come up with had we used our table.

Chapter 45
The Decay of the Positive Tau

> During the decay of the positive tau, the reasons for the creation of both the muon neutrino and the anti-tau neutrino will be examined. These relationships are important because they reveal how the space flowing into a decaying vortex pulls the opposite side of space outward creating an anti-neutrino. It also reveals how a denser region of space pulled towards a flowing vortex rebounds, pulling the opposite side of space inward, creating a muon neutrino.

Figure 45.1 Creation of the tau anti-neutrino $\overline{V_t}$:

As previously explained, the positive tau is a positron containing an up quark that in turn contains a charm quark. This relationship is drawn below:
Note how the positive tau contains three vortices: a 4d vortex, a 5d vortex, and a 6d vortex.
The first vortex, the 4d vortex is the positron vortex. The second, the 5d vortex is the up quark vortex that is associated with the positive muon [designated the MUON VORTEX]. While the 6d vortex is the charm vortex - associated with the positive tau [designated the TAU VORTEX].

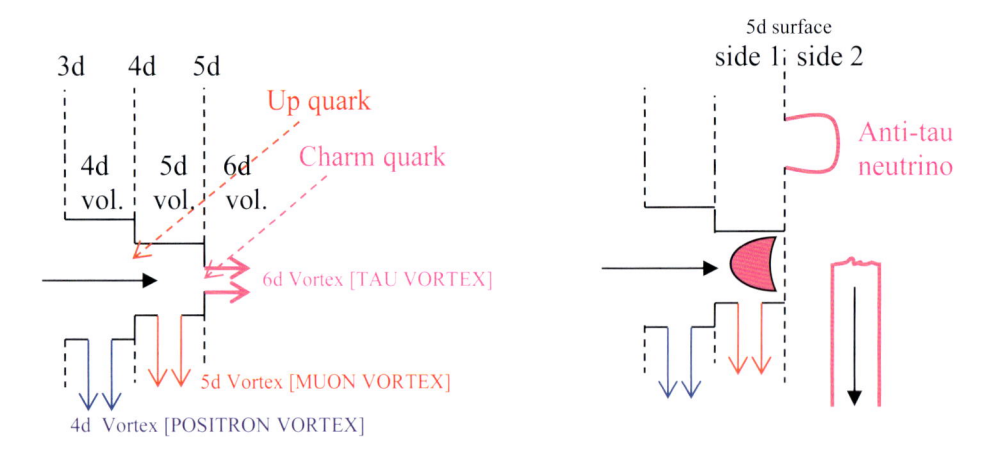

When the charm quark deflates, its collapsing vortex [the 6d tau vortex] pulls side 2 inward. This inward pull creates a volume of space bent into side 2 that becomes the tau anti-neutrino.

THE CREATION OF THE MUON NEUTRINO V_u :

When the charm quark deflates, the extra volume of 5d space within the up quark needed to change the charm quark into a +1 charge deflates too. And again, unlike the decays of the up and down quarks, the decay of the extra volume of 5d space within the up quark takes place in the opposite direction.

Also, like the decays previously seen that create the electron neutrino and the electron anti-neutrino, the reversal of direction occurs because the 5d vortex does not collapse completely. Consequently, the extra volume of 5d space is expelled backward into the 5d volume in the reverse order from which it was collected. Because the outward flow has pulled the extra volume of 5d space into it, the extra volume of 5d space, [shaded area] now pushes back outward into the 5d volume. This sudden outward push from side 2 deforms side 1 inward creating the muon neutrino:

Figure 45.2 Creation of the muon neutrino

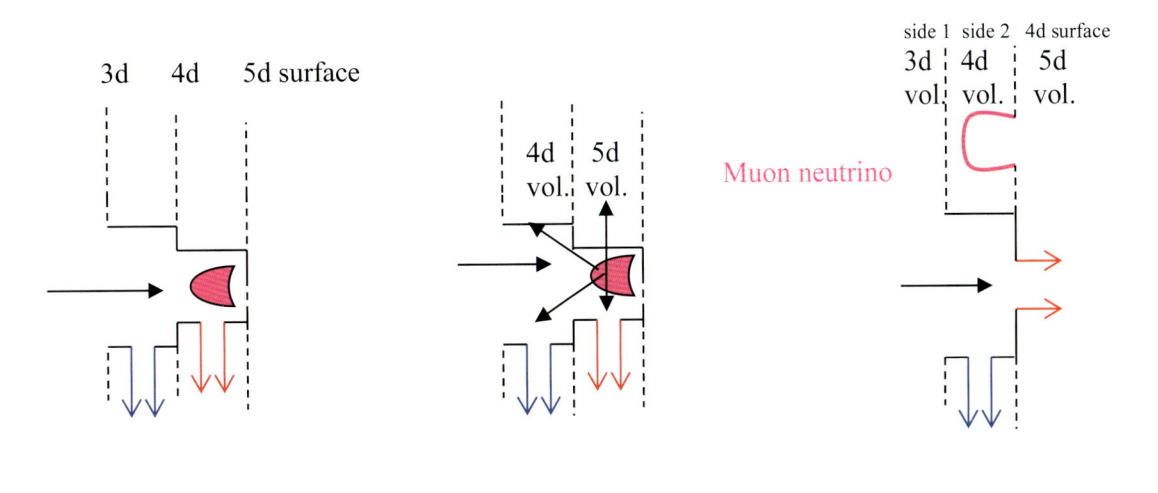

This reaction is represented as the **equation**: $\tau^+ \rightarrow \overline{V_t} + V_u + \mu^+$

CREATION OF CONSERVATION OF LEPTON NUMBER:

In Figure 45.2, note how the direction of the space flowing in the decaying tau vortex pulled the surface of 5d space inward towards side 2. Note too that this direction is the *same* as the direction of the space flowing in the 6d vortex of the positive tau [towards side 2]. This direction of 6d space flowing towards side 2 [previously designated -1] gives the tau anti-neutrino a lepton # of -1; and the direction of the flowing space in the positive tau vortex towards side 2 [previously designated -1] gives the positive tau a lepton # of -1.

Next, in Figure 45.1, [the second half of the decay], notice how the extra volume of 5d space within the up quark, [that was needed to change the +2/3 charge of the charm quark to a value of +1], pushes outward in the opposite direction to that from which it was originally pulled in – creating the muon neutrino. Observe how this outward push is in the opposite direction to the space flowing in the positive tau's vortex. Because this same vortex sheathing the positive tau deflates to become the positive muon's vortex, the direction of the flow in the positive tau's vortex is oriented in a direction opposite to the direction that the positive muon neutrino is oriented. And again, we can now understand and see why a muon neutrino is paired with a positive muon when both are on the same side of the equation.

[Note too, in contrast to the muon neutrino that is formed upon the 4d surface and bent into the 5d volume of space, note how the tau anti-neutrino was created upon the surface of 5d space and is bent into the 6d volume.]

And finally, by just looking at the orientation of how the leptons are bent into the two sides of space, we come up with the following lepton numbers: positive tau = -1; tau anti-neutrino = -1; muon neutrino = +1; and the positive muon = -1. When all four are substituted into the equation representing this decay: $\tau^+ \rightarrow \overline{V_t} + V_u + \mu^+$ the results are: -1 = -1 +1 -1 And we find that we have come up with the exact same result we would have come up with had we used our table.

Chapter 46
The Decay of the Negative Tau

> During the decay of the negative tau, the reasons for the creation of both the tau-anti neutrino and the muon neutrino will be examined. These relationships are important because they reveal how the space flowing out of a decaying vortex pulls the opposite side of space outward creating a neutrino. It also reveals how a denser region of space pushed outwards into the 6d volume, rebounds, pulling the opposite side of space inward, creating a muon anti-neutrino.

CREATION OF THE TAU NEUTRINO

As previously explained, the negative tau is an electron containing an anti-up quark - and the anti-up quark contains an anti-charm quark. This relationship is drawn below:

Figure 46.1 The creation of the tau neutrino V_t :

Note how the negative tau contains three vortices: a 4d vortex, a 5d vortex, and a 6d vortex.
The first vortex, the 4d vortex is the electron vortex. The second, the 5d vortex is the anti-up quark vortex that is associated with the negative muon [designated the MUON VORTEX]. While the 6d vortex is the anti-charm vortex - associated with the negative tau [designated the TAU VORTEX].

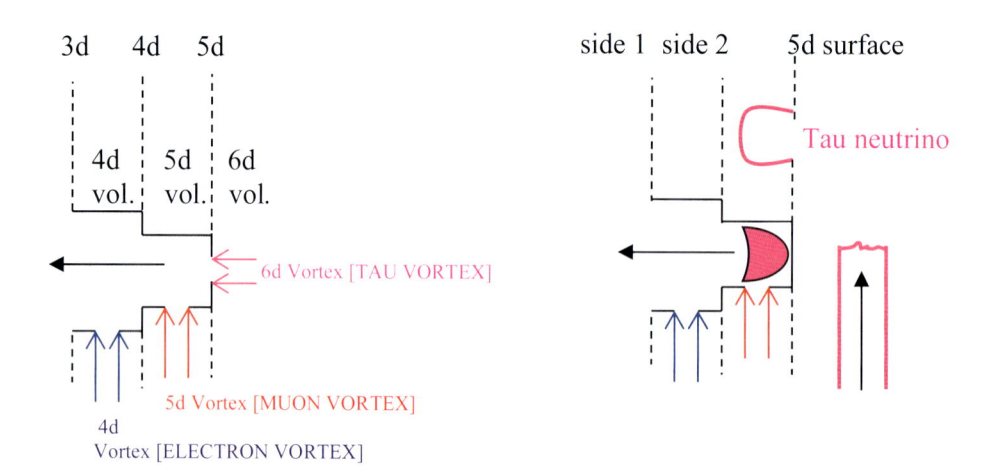

When the negative tau vortex deflates, its inward flow towards side 1, pushes side 1 inward. This inward push creates a volume of space bent into side 1 that becomes the tau neutrino.

Figure 46.2 The creation of the muon anti-neutrino $\overline{V_u}$:

When the anti-charm deflates, the extra volume of 5d space within the anti-up quark needed to change the anti-charm quark into a -1 charge, deflates too. And again, unlike the decays of the up and down quarks, the decay of the extra volume of 5d space within the anti-up quark takes place in the opposite direction.

This reversal occurs because this 5d vortex does not collapse completely. Consequently, the extra volume of 5d space contained within it is expelled in the reverse order in which it was collected. Because the outward flow has pushed the extra volume of 5d space into the vortex, the deflating extra volume of 5d space, [shaded area], now pushes backward upon the 5d volume. This sudden backward push deforms side 2 outward, creating the muon anti-neutrino [Important note: the muon anti-neutrino can be identified by its creation upon the 4d surface of space.]

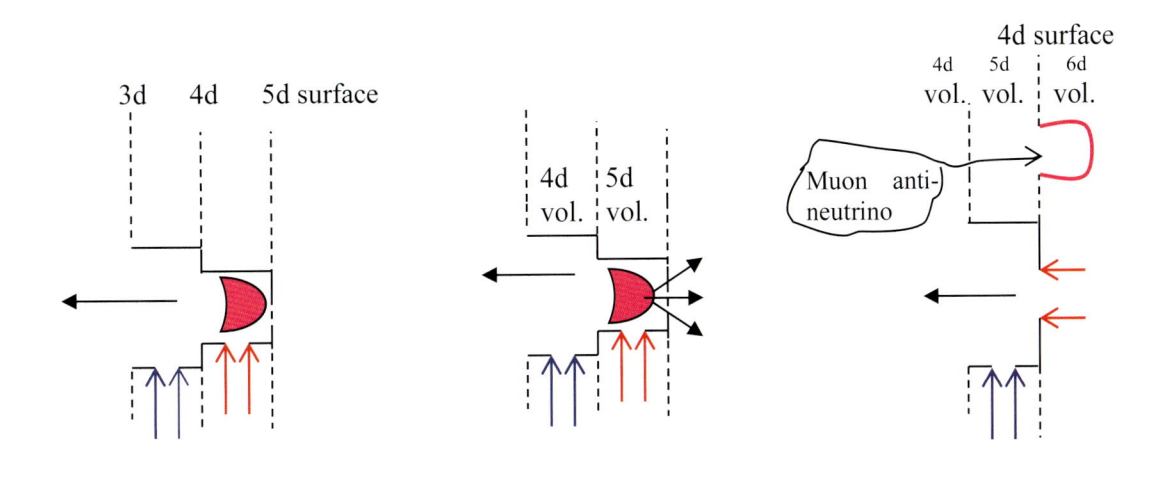

This reaction is represented as the **equation**: $\tau^- \rightarrow V_t + \overline{V_u} + \mu^-$

CREATION OF CONSERVATION OF LEPTON NUMBER:

In Figure 46.1, note how the direction of the space flowing in the decaying negative tau vortex pushed the surface of 5d space outward towards side 1. Note too that this direction is the *same* as the direction of the space flowing in the 4d vortex of the negative tau [towards side 1]. This bend in 4d space towards side 1 [previously designated +1] gives the tau neutrino a lepton # of +1; and the direction of the flowing space in the negative tau vortex towards side 1 [previously designated +1] gives the negative tau a lepton # of +1.

Next, in Figure 46.2, [the second half of the decay], notice how the extra volume of space within the anti-up quark, [that was needed to change the -2/3 charge of the anti-charm quark to a value of -1], now pushes back outward in the opposite direction to which it was originally pushed into the up quark – pushing side 2 outward, creating the muon anti-neutrino. Observe too, how this inward pull is in the opposite direction to the space flowing in the negative tau's vortex. Because this same vortex sheathing the negative tau deflates to become the negative muon's vortex, the direction of the flow in the tau's vortex is oriented in a direction opposite to the direction that the muon anti-neutrino is oriented. And again, we can now understand and see why a muon anti-neutrino is paired with a negative muon when both are on the same side of the equation.

[In contrast to the muon anti-neutrino that is formed upon the 4d surface and bent into the 5d volume of space, note how in Figure 46.1, the tau neutrino was created upon the surface of 5d space and is bent into the 6d volume.]

And finally, by just looking at the orientation of how the leptons are bent into the two sides of space, we come up with the following lepton numbers: negative tau = +1; tau neutrino = +1; muon anti-neutrino = -1; and the negative muon = +1.

When all four are substituted into the equation representing this decay, $\tau^- \rightarrow V_t + \overline{V_u} + \mu^-$

and the results are: +1 = +1 -1 +1 And just as before, we find that we have come up with the exact same result we would have come up with had we used our table.

In the next section, two final examples of neutrinos and the products they create are given:

Chapter 47
The Decay of the "WOW" Lepton

The WOW lepton is a new particle proposed in nature. Its innovation is a result of the number of dimensions proposed in the Vortex Theory.

The signature decay of the two oppositely charged "Wow" lepton's will be seen in: the mutual creation of two oppositely charged Tau's they decay into; the odd angles the Tau's travel at in relation to each other; and the odd angles of the muons the Tau's subsequently decay into.

When two Wows' are mutually created in a particle collision, they will move away in opposite directions to each other. They will be invisible because of their short lifetime. However, what will indicate the presence of the Wow leptons will be the odd angular path of its decay products.

These odd angular decay paths will be created by the presence of the Wow neutrino and its interaction with the Tau decay product. The Wow neutrino will cause the Tau to veer off at an angle rather than in a straight line. Each Tau on either side of the creation point will possess the same angular deviation from a line drawn tangent to the point of creation during the impact of the primary "particles" in the particle collider.

It is theorized that the negative Wow will decay as follows:

Wow → Wow Neutrino + Tau anti-neutrino + negative Tau

Or: $W_0^- \rightarrow V_{wow} + \overline{V_t} + \tau^-$

It is also theorized that the negative Wow will decay into negative B mesons, especially the negative B charm and the negative B up.

$$W_0^- \rightarrow B_c^-$$

$$W_0^- \rightarrow B_u^-$$

Because the decay goes from side 2 to side 1, either a charm or an up quark from side 2 is expected to survive the decay. Since the strange and the down are side 1 quarks, it is not expected to see the B strange or the B down.

Chapter 48
The Collision Between a Proton and an Electron Anti-Neutrino
[And: The Proton, Muon Anti-neutrino Collision]

One of the most famous reactions in particle science is the collision between a proton and an anti-neutrino. When the proton and anti-neutrino collide, they produce a neutron and a positron. This collision can now be explained using our understanding of the two sides of space and the way anti-neutrinos are created.

Figure 48.1 Collision between a proton and an anti-neutrino

In the drawings below, the anti-neutrino approaches and then strikes the proton:

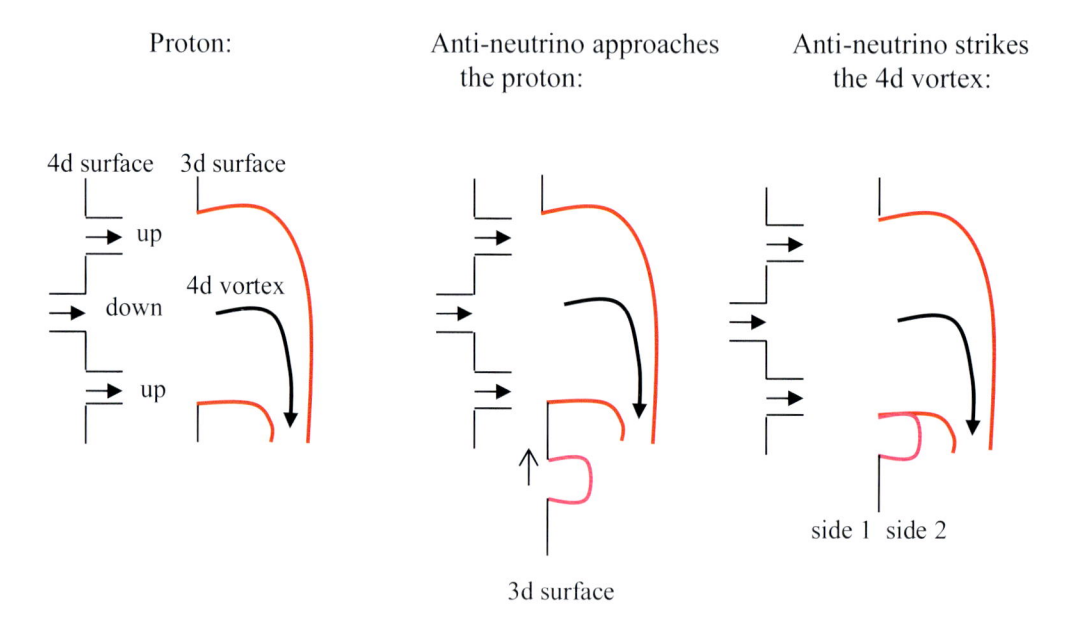

Moving at the speed of light, when the anti-neutrino strikes the 4d vortex, *its volume of space bent into side 2 is suddenly added* to the 4d vortex's volume of flowing space that is also bent into side 2. The almost instantaneous inflation causes three simultaneous events to occur:

#1. The 4d vortex breaks. One end curls back upon itself creating a 4d torus; while the other end continues to flow into side 2 becoming the positron:

Figure 48.2

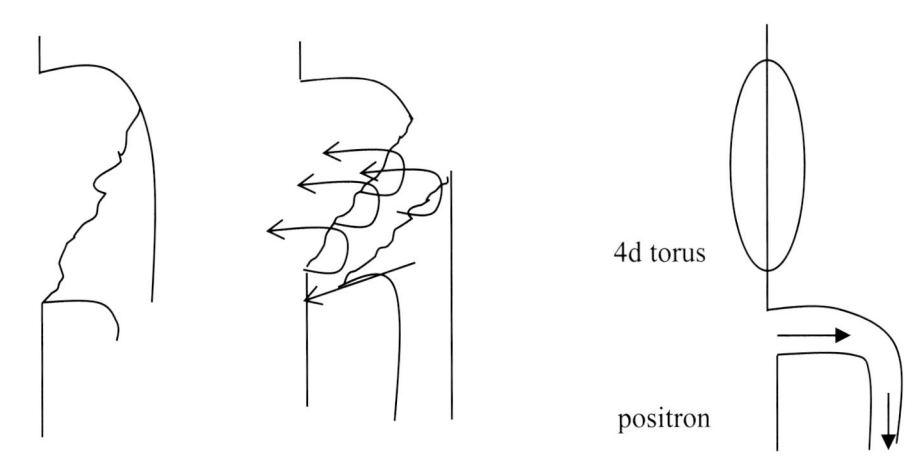

4d torus

positron

#2. The sudden outward distortion of the space inside the proton by the anti-neutrino causes the up quark closest to the incoming anti-neutrino to suddenly increase in size. This sudden increase in size exceeds the elasticity of the space it is formed in and as soon as it increases in size, it collapses. As it collapses, it pulls side 1 outward towards side 2 creating a pair of down, anti-down quarks. As the down, anti-down quarks are being created, and the breaking 4d vortex forms a torus, a circulating flow within the newly forming neutron, both the anti-down and the up quarks collapse via the "breakaway decay with neutrino" mode.

Note: it appears as if the anti-down quark and the up quark should unite to form a positive pion inside the vortex created by the positron. So, why doesn't it? It appears as if there is not enough 4d volume present. If there were, the size of the positron's vortex would increase allowing the anti-down and the up quark to create a positive pion.

#3. When the up quark deflated, creating the down quark, the outward flowing space at the other end of the down vortex, [the anti-down quark], suddenly subtracts from the volume of 4d space contained within the 4d vortex. Because this creates an imbalance between the 4d vortex and the volume of 4d space contained within it, the anti-down quark is expelled from the neutron; so upon its formation, the anti-down quark immediately deflates and tunnels through 4d space. As it deflates, its inward flowing space causes side 1 to pull inward, creating the <u>effect</u> of a muon neutrino. However, because the outward flowing space of the deflating up quark pulls side 2 inward, creating the <u>effect</u> of a muon anti-neutrino, as seen from the 3d surface, *the two opposite effects cancel each other, and no neutrino is seen. In Figure 48.4, all that is left is the neutron and the positron:*

Figure 48.3 **Figure 48.4**

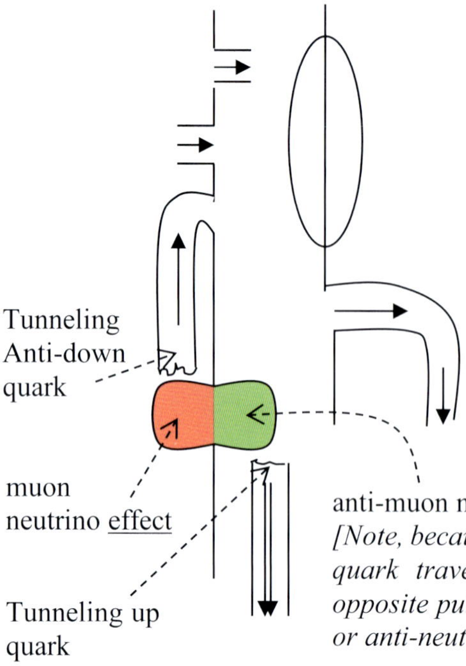

Tunneling Anti-down quark

muon neutrino <u>effect</u>

Tunneling up quark

anti-muon neutrino <u>effect</u>
[Note, because the tunneling anti-down quark and the tunneling up quark travel side by side simultaneously, their simultaneous opposite pulls upon the surfaces of space cancel and no neutrino or anti-neutrino is created: only their effects are felt..]

Neutron

Positron

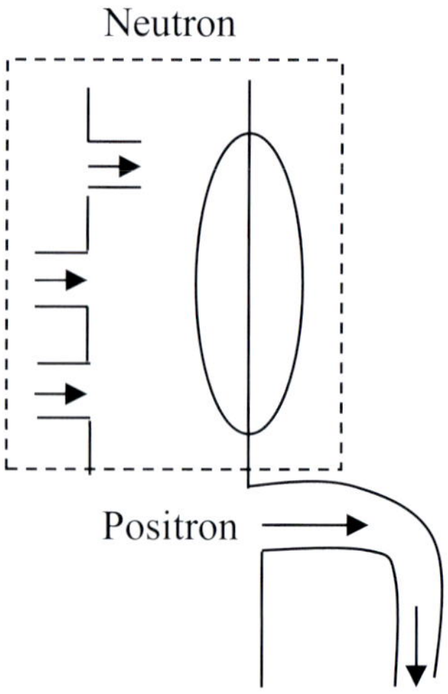

Also, as the tunneling down quark peels off a layer of side 2 [pulling side 2 into side 1], the tunneling up quark pulls side 1 back into side 2 sealing the rift creating a quiescent surface.

PROTON, MUON ANTI-NEUTRINO COLLISION

When the muon anti-neutrino containing a <u>tunneling up quark</u> strikes the proton, almost the exact same set of circumstances is created. However, because the up quark in the muon anti-neutrino contains a volume of 5d space that displaces the 4d volume within the proton, when the up quark is suddenly added to the proton, the volume of the proton is further increased, increasing the size of the positron at the other end; allowing the up quark to stay within the positron – forming the positive muon.

When the muon anti-neutrino containing a tunneling down quark strikes the proton, the down quark's 4d volume adds to the space of the proton causing it to form the neutron; the down quark adds to the forming neutron; and the up quark adds to the forming positron, increasing its size, forming the positive muon.

ANTI-PROTON COLLISION WITH A POSITRON NEUTRINO

For an anti-proton positron collision, the same exact explanation is discovered as explained in 48.1. However, the anti-proton is now constructed out of two anti-up quarks & an anti-down quark; and the positron neutrino is bent outward into side 1.

ANTI-PROTON COLLISION WITH AN ANTI-MUON NEUTRINO

Again, the explanation is the same as in 48.2, except that the anti-muon neutrino contains a tunneling anti-down quark.

Chapter 49
The Collision Between a Neutron and an Electron Neutrino; And the Decay of the Neutron and the Creation of the Anti-neutrino

This chapter is actually two chapters in one, it contains the explanation of the collision between the neutron and the electron neutrino, and the explanation of the neutron decay and the creation of the anti-neutrino. The collision between the neutron and the electron neutrino is discussed first.

THE COLLISION BETWEEN A NEUTRON AND AN ELECTRON NEUTRINO

When an electron anti-neutrino collides with a neutron, the reverse set of events to what we have just seen occurs:

Figure 49.1 the electron neutrino strikes the neutron:

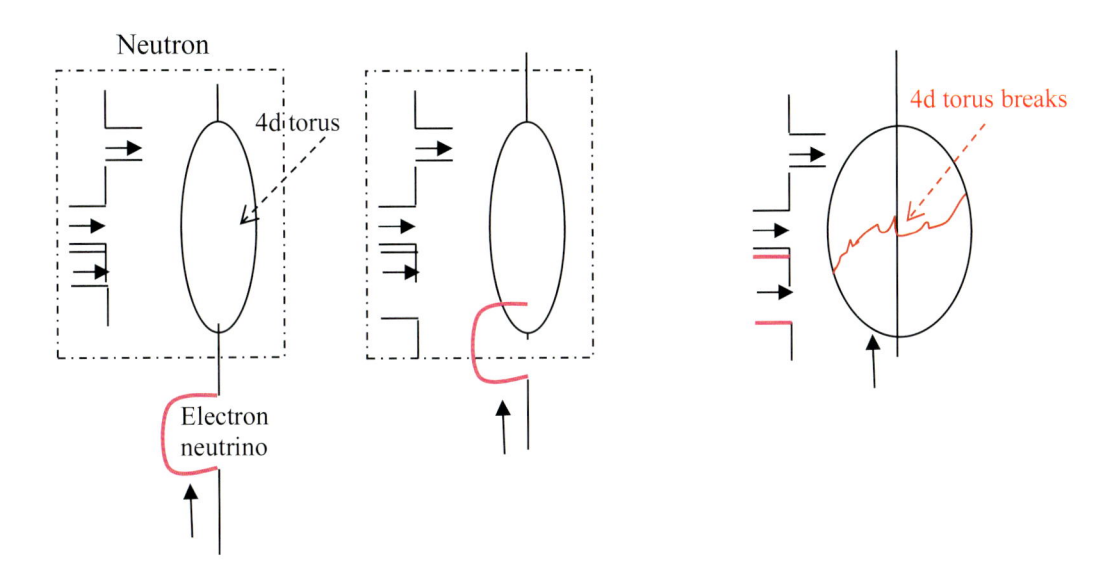

Just as the electron anti-neutrino increased the volume of the proton creating the chain of events that turned it into the neutron, the electron neutrino decreases the volume within the neutron creating the chain of events that allows it to turn back into a proton:

As the electron neutrino collides with the neutron, its volume of 4d space bent into side 1 suddenly subtracts from the volume of 4d space that is bent into side 2. Because the 3d surface of the neutrino is bent outward into side 1, when it impacts into the 3d surface of the 4d torus, a tear or rip is created upon the 3d surface of the torus. This tear causes the torus to break.

At the same time, because the neutron is moving at the speed of light, when it impacts into the neutron, its subtraction from the volume of space within the neutron occurs at the speed of light. This almost instantaneous decrease in the volume of the space within the neutron suddenly increases the size of the down quark nearest to the point of impact [note, the down quarks allow space to flow into the volume within the neutron].

Then as the torus breaks and begins to reform the two ends of the 4d vortex, the increased size of the down quark contracts causing side 2 to bulge outward creating the up anti-up quark pair. Because the outward flowing space at the other end of the up vortex, [the anti-down quark], adds to the volume of the proton, it cannot stay within the reforming proton. Instead, it immediately deflates and tunnels through 4d space. As it deflates, its outward flowing space causes side 1 to push outward, creating the <u>effect</u> of a muon neutrino. However, and at the same time, because the down quark has also deflated, its outward push into side 2 creates the <u>effect</u> of a muon anti-neutrino. These two opposite effects cancel each other, and no neutrino is seen. All that is left is the proton and the electron seen below:

Figure 49.2 as the torus breaks…

[Note, the breaking 4d torus creates the W⁻ Particle.]

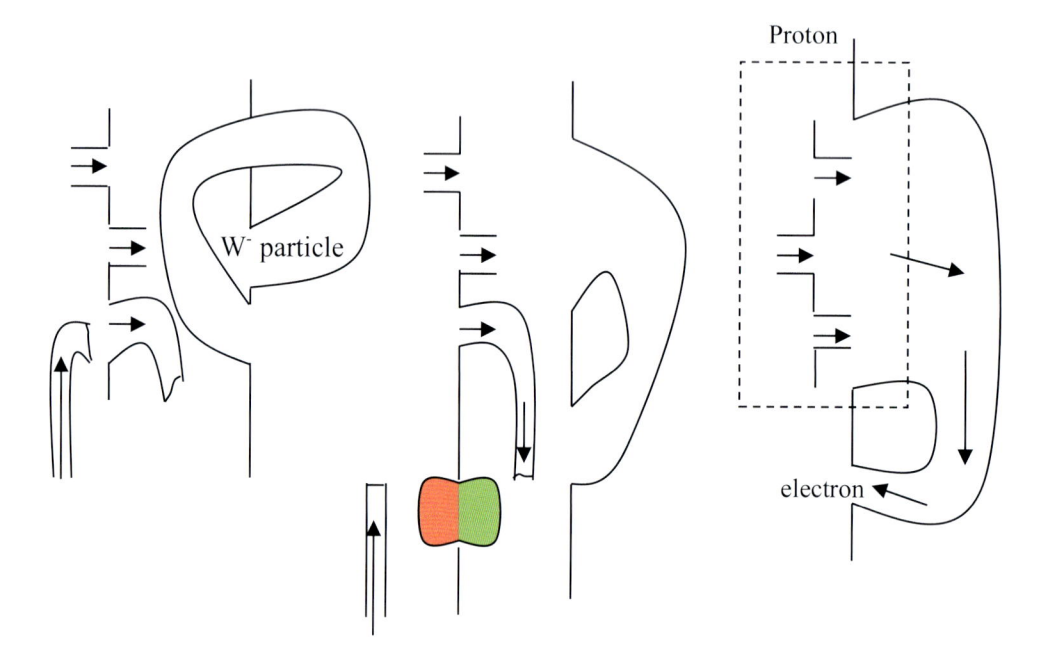

[As before, because the tunneling anti-down quark and the tunneling up quark travel side by side simultaneously, their simultaneous opposite pulls upon the two surfaces of space cancel and no neutrino or anti-neutrino is created: only their effects are felt.]

NEUTRON MUON NEUTRINO COLLISION

When the muon neutrino containing a tunneling anti-up quark strikes the neutron, almost the same set of circumstances is created. The muon neutrino's volume of 4d space adds to the volume of one of the down quarks causing it to decay – pulling side 2 outward – creating the up quark, and the anti-up tunneling quark. But also, since the muon neutrino contains a volume of 5d space, when the anti-up is suddenly added to the neutron, the volume of the neutron increases and adds to the volume of the forming electron, the anti-up is pulled into this space and a negative muon is created.

When the neutron is struck by a muon neutrino containing a tunneling anti-down quark, a completely different set of circumstances is created. The anti-down quark and one of the neutron's down quarks annihilate creating an up, anti-up pair of quarks. The up combines with the other up and the other down to form the existing proton, while the anti-down combines with the forming electron to produce the negative muon.

Chapter 50
The Decay of the Neutron and the Creation of the Anti-neutrino

The decay of the neutron is very similar to the collision between a neutron and a neutrino: the vortex breaks and flips from inside out to right side out forming what science believes to be a "W" particle. As one end of the decaying vortex again becomes the proton and the other becomes the electron, the vortex is no longer a torus. As all of this is happening, the down quark decays and tunnels forming an up quark and a tunneling anti-up quark; the tunneling down quark and the anti-up quarks tunnel off together, the 4d effects of one feeds the 4d volume needs of the other vortex while negating the 3d surface effects of the other – creating a quiescent surface. After the vortex is turned right side out, an anti-neutrino is created, and this is where the distinction in the differences in the collision and the decay commence.

When a neutron is originally created, a small amount of space equal to the volume contained within an anti-neutrino is added to its 4d volume. If the neutron anti-neutron are created simultaneously in a collision, amid the violence and chaos of the collision this effect is not seen. Only its effect is seen when the neutrino collides with a neutron and its outward bent volume of 4d space suddenly cancels the inward bend of the anti-neutrino, canceling its volume addition to the neutron causing the neutron to decay. [This ejection of 4d volume is also seen when the neutron decays on its own and this extra volume of space is thrown from its volume in the form of an anti-neutrino.]

This injection or ejection of additional volume is important because the balance between the charges on the positive and negative quarks only allow the volume of space flowing into the fourth dimensional volume within the three dimensional hole to remain constant. They have nothing to do with the space flowing in the vortex. The loop created by the vortex does not allow additional 3d space to flow into or out of the vortex.

If additional space does flow in, the vortex breaks. This situation is created when the neutrino strikes the neutron. The neutrino would add just enough 4d volume to the 3d hole, and enough 3d space [the outside of the neutron is constructed out of 3d space] to the twisted vortex to force the vortex to expand, causing it to break. This breaking vortex then allows 3d space to again flow into the three dimensional hole.

The resumption of the flow of three dimensional space into the 3d hole expands the size of the hole. The expansion of the hole forces the volume of 4d space within the hole to increase in volume. This increase in volume forces the quark that is keeping the flow constant to deflate. Its deflation opens up a new hole in the opposite side of space.

This new hole that opens up is similar to a hole opening in the wall of a dam, this hole has to open up in the direction that allows space to now flow out of the hole. So, if space is flowing into the hole giving it a positive charge, the hole that opens up has to also possess a positive charge. This means that the 4d hole within that had a negative charge and allowed 4d space to flow into the hole has to close up so that space can now flow out in the opposite direction. Within a neutron, the hole that has to open up is an up quark and the hole that has to close down is a down quark.

Also, [as before], because the tunneling anti-down quark and the tunneling up quark travel side by side simultaneously, their simultaneous opposite pulls upon the two surfaces of space cancel and no neutrino or anti-neutrino is created: only their effects are felt, and no additional particles are seen. Only the proton, the electron and the electron anti-neutrino appear to be created.

However, it is most fascinating to note that if during the construction of the neutron, more 4d space was somehow added to the interior of the neutron, its torus would be bigger creating a most unusual situation. The added 4d space would allow more space for additional quarks. Consequently, instead of tunneling, the down quark and the anti-up quark would form within the electron creating a negative pion. When expressed in terms of nucleons the results are: (N π^-)! And this appears to be exactly what is happening.

In the CRC *Handbook of Chemistry and Physics*, in the Baryon Summary Table, it can easily be seen that when a nucleon (N) decays, if due to the spins of the quarks, it possesses a higher energy state, then it wants to decay into a nucleon (N) and a pion (π)! Confirming our original hypothesis!

Because a nucleon consists of protons and neutrons, an added confirmation can be seen when an energetic proton is turned into a neutron (N) and a positive pion (π^+). Instead of a positron being created, the greater volume of 4d space within the positron would allow the up quark and anti-down quark to remain within this 4d volume instead of tunneling. They would combine with the positron to form the positive pion: (N π^+ expressed in terms of nucleons) Again, quite a remarkable hypothesis!

Chapter 51
The Stability of the Proton;
the Instability of Mesons

The stability of the proton begins with the realization that like protons and electrons, quarks are also surrounded by regions of dense and less dense regions of space. Just as protons and electrons possess spherical regions of 3d space bent into and out of them, quarks possess spherical regions of 4d space bent into and out of them:

The up quark: because the up quark's charge is +2/3, 4d space is flowing into it. Because space is flowing into it, it is surrounded by a less dense region of space. Since the up quark is of side 1 construction, it is pictured green.

Because the anti-up quark's charge is -2/3, 4d space is flowing out of it. It is also green because of its side 2 construction:

Figure 51.1

UP QUARK   ANTI-UP QUARK

Because the charm and top quarks are contained within up quarks, they will also be similar to the up quark; also, because the anti-charm and anti-top quarks are also constructed out of anti-up quarks, they will be similar to the anti-up quark.

The down quark: because the down quark's charge is -1/3, 4d space is flowing out of it. Because space is flowing out of it, it is surrounded by a denser region of space. Since the up quark is of side 2 construction, it is pictured red.

Because the anti-down quark's charge is +1/3, 4d space is flowing into it. It is also side 2 construction it is also pictured as red:

Figure 51.2

ANTI- DOWN QUARK   DOWN QUARK

Because the strange and bottom quarks are contained within down quarks, they will also be similar to the down quark; also, because the anti-strange and anti-bottom quarks are also constructed out of anti-down quarks, they will be similar to the anti-down quark.

THE STABILITY OF THE PROTON

The stability of the proton is no accident of nature. This stability is a function of the dense and less dense regions of space created by the proton's two up quarks and one down quark.

Because the two up quarks with their positive charges possess volumes of space bent into them, the down quark possesses a volume of space bent out of it, when all three are together within the proton they create a symmetrical "spatial" relationship.

This symmetrical spatial relationship is created with the down quark in the middle and the two up quarks rotating around it. When all three are in this configuration, the dense region of space created by the down quarks is equally balanced on either side by the less dense regions of the up quarks. Having the less dense regions on either side of the down quark, causes its 4d surface to distort outward towards both up quarks simultaneously changing it into a football shape. At the same time, although the surfaces of the up quarks want to distort into pear shapes pointing them and pulling them towards the down quark in the middle of the proton, they cannot.

They cannot because the bent outward space surrounding the down quark is twice as dense as the space surrounding the up quarks. Because this space is twice as dense, the up quarks move outward away from the down quark to a position where the space flowing out of the down quark and flowing into the up quark are the same density. These densities and their corresponding shapes cause the down quark to remain positioned in the center of the proton with the quarks directly opposite to each other: creating a stable configuration. This stable configuration keeps the similar charges of the up quarks from accelerating away from each other - breaking up the proton.

THE INSTABILITY OF MESONS

The reason why mesons are unstable particles comes from the observation that to possess a positive charge, they have to have a quark with a +2/3 charge and a quark with a +1/3 charge. However, because these quarks are both surrounded by regions of bent inward space, they repulse each other. Because there is no third quark to interfere with their repulsion, the more massive quark falls prey to the elasticity of space that wants to contract to alleviate this stressful bend. Hence it contracts causing the more massive quark to decay.

The exact same scenario is true for the negatively charged mesons with their -2/3, and -1/3 quarks.

PART IV
THE SIX CONSERVATION LAWS OF NATURE FINALLY EXPLAINED

Chapter 52
Explanation of the Conservation Law of Lepton Number

HOW THE LAW OF THE CONSERVATION OF LEPTON NUMBER WAS DISCOVERED

The law of the conservation of lepton number is explained by the two sides of space. The compression of one side against the other as particles decay creates new particles on the other side.

The explanation of the Law of the Conservation of Lepton Number was discovered when the Principles of the Vortex Theory of Atomic Particles were applied to the Quark Theory.

When the Vortex Theory was first discovered, it was theorized that the red shift of the Galaxies could be explained if the universe was constructed out of a single giant expanding higher dimensional sphere. Just as a two dimensional plane is the surface of a three dimensional sphere, the three dimensional space of our universe would be the surface of this fourth dimension volume. Charged particles would exist as three dimensional holes upon its surface, the forces of nature would be constructed out of bent and flowing space, photons would be quantized compression waves traveling through this surface, and neutrinos would be quantized transverse waves bent into and out of the three dimensional surface – like ripples upon the surface of a vast ocean. The isotropic spin of particles could be explained by vectors pointing into and out of higher dimensional space. This model was most successful.

Using this model, *all* of the phenomena associated with the Theory of Relativity were successfully explained, as were the great mysteries of classical physics [such as Newton's Three Laws of Motion]. Unfortunately, most of the phenomena associated with quantum mechanics could not be explained; nor could the phenomena associated with the Quark Theory be explained, nor was this model capable of explaining what this giant ball of space was expanding into. At the time of the conception of the original theory it was believed that this expanding sphere of space was expanding into nothing: a void. But this view changed dramatically when the theory was finally applied to quarks.

When the Vortex Theory was finally applied to quarks, it became evident that the expanding volume of space was expanding into another volume of space that was contracting. These two volumes of space sharing a mutual surface create a unique situation: two sides of space. The surface of the expanding volume can be called Side 1, and the surface of the contracting volume can be called Side 2.

It also became clear that if this second volume of space possesses one fourth the density of the first volume, then the $\pm 1/3$ and $\pm 2/3$ charges of the quarks could also be explained. Also, like the baryons and mesons they exist within, quarks are easily explained if they too are holes in higher dimensional space connected to their anti-particles by vortices in still higher dimensions.

Like the layers of a cake, quarks too are layered; up quarks [formed on Side 1] and down quarks [formed on Side 2] are fourth dimensional holes existing within the fourth dimensional volume of space within three dimensional mesons and baryons; strange and charm quarks are fifth dimensional holes existing within the fifth dimensional volume of space within up and down quarks; while top and bottom quarks are sixth dimensional holes existing within the sixth dimensional volume of space within strange and charm quarks.

The explanation of neutrinos also underwent a remarkable transformation. It was discovered that there are two types of neutrinos: those that are pure quantized transverse waves; and those transverse waves that are accompanied by a "tunneling" quark.

The word "tunneling" was used due to the similarity between it and the transformation of the electron in the tunneling diode. When applying the principles of the Vortex Theory to the mystery of the tunneling diode, it was discovered that the electron is not turned into a "wave" as is presently believed. Instead, when the three dimensional hole we call the electron reaches the junction barrier, the 3d hole deflates, the electron vortex travels through 4d space "beneath" the three dimensional junction barrier at the speed of the space flowing in it (i.e. the speed of light); then once past the barrier, the end of the vortex jumps back into 3d space to again become the three dimensional hole we call the electron. This same effect occurs when a quark seems to decay creating a neutrino.

Using the up quark as an example, because the up quark is really just a 4d hole that is one end of a 5d vortex; when the up quark tunnels, the hole is pulled into the 5d volume, and this end of the vortex elongates outward at the speed of the vortex (i.e., the speed of light).

Because space is flowing into the up quark, as this end of the vortex tunnels, its actions are analogous to it pulling the 4d surface down into it, creating a trough or furrow in 4d space that dissipates out into the surrounding 4d surface as it passes. From our point of view, the wave front of this trough is perceived to be the anti-neutrino.

When the anti-up quark tunnels, because space is flowing out of it, the exact reverse happens. In this case, a ridge or thin ribbon of space is laid down that dissipates into the surrounding space. The wave front of this structure is perceived by us – from our point of view – to be the neutrino.

All of the appropriate background information necessary to prepare the reader to understand the Law of the Conservation of Lepton Number has now been provided, and our discussion can begin.

A BRIEF REVIEW OF THE LAW OF THE CONSERVATION OF LEPTON NUMBER

The foundation of the law of the conservation of lepton number is based upon the observation that during particle collisions or decays, certain leptons are either created or destroyed together. The following sets of rules governing what you start out with and what you end up with have been painstakingly discovered and carefully documented by particle scientists. These rules are reviewed here:

TABLE 52-1 EIGHT LEPTON RULES

When leptons are on the *same side* of the reaction equation the following rules are used:
1. positive muons are accompanied by neutrinos.
2. negative muons are accompanied by anti-neutrinos.
3. positrons are accompanied by neutrinos.
4. electrons are accompanied by anti-neutrinos.

When leptons are on the *opposite side* of the reaction equation these set of rules are used:
5. positrons are accompanied by anti-neutrinos.
6. electrons are accompanied by neutrinos.
7. positive muons are accompanied by anti-neutrinos.
8. negative muons are accompanied by neutrinos.

To explain the above mentioned rules mathematically, each lepton is assigned a set of numbers listed below in Table 52-2:

TABLE 52-2 LEPTON NUMBERS

LEPTON	ELECTRON #	MUON #	TAU #
Positron	-1	0	0
Electron	+1	0	0
Electron anti-neutrino	-1	0	0
Electron Neutrino	+1	0	0
Positive muon	0	-1	0
Negative muon	0	+1	0
Muon anti-neutrino	0	-1	0
Muon neutrino	0	+1	0
Positive tau	0	0	-1
Negative tau	0	0	+1
Tau anti-neutrino	0	0	-1
Tau neutrino	0	0	+1

Looking at Table 52-2, it can be seen that the Lepton Number for each lepton consists of *three* different numbers: an electron #, a muon #, and a tau #. Because all three of these numbers must be conserved *together*, only an electron # can accompany an electron #; only a muon # can accompany a muon #; and only a tau # can accompany a tau #.

So why do these relationships occur? We begin our discussion by investigating three relationships: The first is called "the two types of leptons":

THE TWO TYPES OF LEPTONS

Although present day science arranges all leptons into one category, this is a mistake. There are actually two types of leptons: "type 1", those that are holes with space flowing through them; and "type 2", those that are indentations in space.

Type 1: <u>Charged Leptons</u>: electrons, positrons, muons, anti-muons, taus, and anti-taus are the ends of vortices of flowing space. As such they are holes in space.

Type 2: <u>Neutral Leptons</u>: neutrinos and anti-neutrinos are partial volumes of these holes that are expelled into space. As such, their appearance is similar to an indentation: a quantized bulge or "bubble" bent into or out of the surface of space. These bends occur upon the surfaces of different dimensions of space: electron neutrinos and electron anti-neutrinos are formed upon the surface of 3d space and bent into and out of the 4d volume; muon neutrinos and muon anti-neutrinos are formed upon the surface of 4d space and bent into and out of the 5d volume; while tau neutrinos and tau anti-neutrinos are formed upon the surface of 5d space and are bent into and out of the 6d volume. [However, the muon neutrino's 5d bend in space continues down onto the surface of 4d space giving it a 4d component; while the tau neutrino's 6d bend continues down onto 5d space and then 4d space giving it a 5d as well as a 4d component.]

TWO TYPES OF TYPE 2

Although science is unaware of it, there has to be a second type of neutrino and anti-neutrino. This second type is a quantized wave accompanied by a "tunneling" quark. [The effects of the second type will be revealed in our discussion.] For now, it should be said that this tunneling effect is mimicked by the actions of the electron in the tunneling diode as previously mentioned.

THREE IMPORTANT CRITERIA

The reason why certain type 2 and type 1 leptons are associated with each other is a direct result of three important criterion: #1 the dimension a vortex is formed upon; #2 the side of space a vortex exists upon [side 1 or side 2]; and #3 the direction space is flowing within a vortex; These three important principles determine what type of bend in space will be created. This bend then determines what type of neutrino or anti-neutrino will be created. *[Note: the anti-symmetric parity of neutrinos is a function of the physical characteristics of the two sides of space and is independent of the particles that created them. (The creation of anti-symmetric parity is explained by the principles of the Vortex Theory, but further experimentation is needed.)]* We will see the results of these three important criteria throughout the rest of our discussion.

LEPTON ORIENTATION IN SPACE

A common characteristic that type 1 and type 2 leptons share is their orientation in space. Even though type 1 and type 2 leptons are structured differently they nevertheless possess distinct orientations in space. For example, neutrinos are all bent into side 1 of space, while anti-neutrinos are all bent into side 2. Electrons, negative muons, and negative taus are all holes constructed on side 2 of space who's *outward flowing space flows into side 1*; while positrons, positive muons, and positive taus are all holes constructed on side 2 whose *inward flowing space flows into side 2*.

In looking at the above orientations, it is important to note that all leptons point in either one of two directions. If positive and negative numbers are assigned to these two directions, and if the direction towards side 1 is called +1 and the direction towards side 2 is called -1, all of the lepton numbers in table 52-1 can be extrapolated.

Figure 52.1. Lepton orientations in space:

If the direction pointing towards side 1 of space is given a value of +1, and the direction pointing towards side 2 is given a value of -1, the numbers of the twelve different leptons can be determined by the direction of the side they bend or flow into. In the below drawings, notice how the direction of the flowing space determines what the lepton number is:

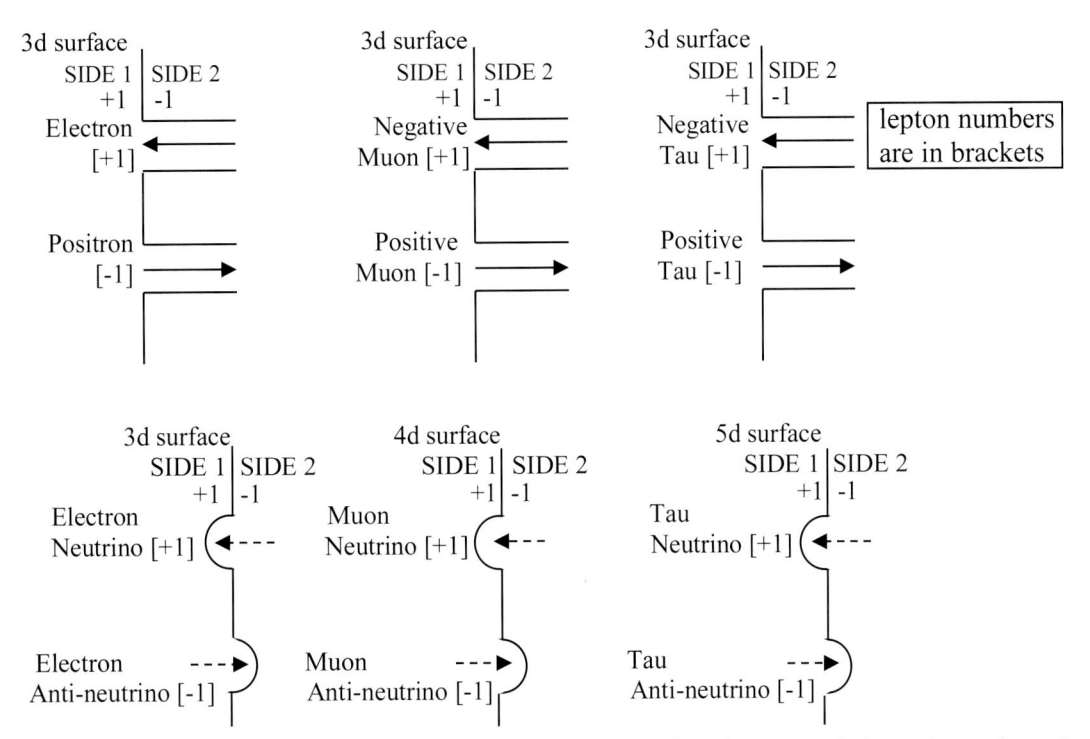

```
    3d surface                3d surface                3d surface
  SIDE 1 | SIDE 2           SIDE 1 | SIDE 2           SIDE 1 | SIDE 2
    +1   |  -1                +1   |  -1                +1   |  -1
  Electron                  Negative                  Negative           ┌──────────────┐
   [+1]     ◄────            Muon [+1] ◄────          Tau [+1] ◄────     │lepton numbers│
                                                                         │are in brackets│
                                                                         └──────────────┘
  Positron                  Positive                  Positive
   [-1]     ────►            Muon [-1] ────►          Tau [-1] ────►
```

```
    3d surface                4d surface                5d surface
  SIDE 1 | SIDE 2           SIDE 1 | SIDE 2           SIDE 1 | SIDE 2
    +1   |  -1                +1   |  -1                +1   |  -1
  Electron                  Muon                      Tau
  Neutrino [+1] ◄---        Neutrino [+1] ◄---        Neutrino [+1] ◄---

  Electron     ---►         Muon        ---►          Tau         ---►
  Anti-neutrino [-1]        Anti-neutrino [-1]        Anti-neutrino [-1]
```

Observing the above diagrams, it can now be seen that the direction space is bent determines the lepton number. Observe too that the electron, muon, and tau neutrinos, [and their anti-neutrinos] are all created upon the surfaces of different dimensions. These dimensions correspond to the higher dimensional holes the Vortex Theory now proposes to exist within the muon, tau, and their anti-particles. The creation of these Neutrino, Anti-neutrino bends is discussed. One example will be given for each of the eight lepton rules.

THE EXPLANATION OF RULE #1 & #2 FROM TABLE 52-1: *why neutrinos are accompanied by positive muons; and why anti-neutrinos are accompanied by negative muons:*

The reason why the muon neutrino is created during the decay of the positive pion, and the anti-muon neutrino is created during the decay of the negative pion will be examined first. These relationships are important because they reveal how the space flowing in a vortex pushes or pulls against the side of space opposite to it when it decays. It is this pull or push that bends the opposite side of space inward or outward creating the muon neutrino or muon anti-neutrino.

Figure 52.2 The decay of the positive pion <u>creates a muon neutrino</u>: V_u

In the below drawings, the decay of the anti-down quark is shown during the decay of a positive pion π^+. Note how the positive pion is constructed out of an Up quark and an Anti-down quark. [The curved lines containing the 4d vortex represents the expanded positron vortex the positive muon is constructed out of.]

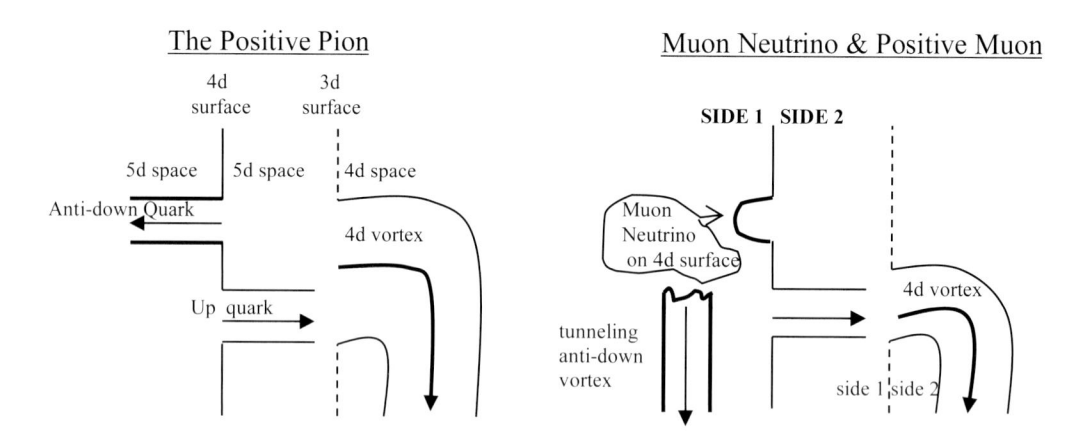

When the anti-down quark deflates, its inward pull [note the direction of space (towards side 1) flowing in the anti-down quark] into side 1 pulls side 2 outward. This inward pull creates a quantized volume of 5d space bent into side 1 that becomes the Muon Neutrino. The anti-down vortex "tunnels" through space with the muon neutrino

The removal of the 5d volume of space previously occupied by the decayed anti-down quark causes the 4d surface of the 5d volume [the 4d volume within the pion] to be reduced in size, forcing the 4d vortex to reduce in size creating the positive muon.

This reaction is represented as <u>**equation 1**</u>: $\pi^+ \rightarrow V_u + \mu^+$

CREATION OF CONSERVATION OF LEPTON NUMBER:

In Figure 52.2, note that the deflating anti-down quark pulled the surface of 4d space inwards, towards side 1, in an opposite direction to that of the space flowing in the 4d vortex of the newly formed positive muon. This bend in 4d space *towards side 1*, [previously designated +1], gives the muon neutrino a lepton # of +1; and the direction of the flowing space in the positive muon vortex *towards side 2* [previously designated -1], gives the positive muon a lepton # of -1. Because the positive pion has a lepton # of 0, when all three numbers are plugged into <u>equation 1</u>, **0 = +1 -1**, we see that the conservation of lepton number is conserved. But even more important, for the first time we suddenly begin to see how the conservation of lepton number is being generated!

  (We have also just seen one of the reasons why a muon neutrino is created along with the creation of a positive muon. Note: the anti-down quark uses the neutrino to pull a double portion of 4d space out of side 2 as it tunnels; allowing it to maintain its flowing volume in one direction as it tunnels in the opposite direction.

Figure 52.3 The decay of the negative pion:

In a negative Pion, the decay of the down quark creates a muon anti-neutrino $\overline{V}_u$.
In the below drawings, the decay of the down quark is shown during the decay of the negative pion π-.

Note too: the negative pion is constructed out of a down quark and an anti-up quark.

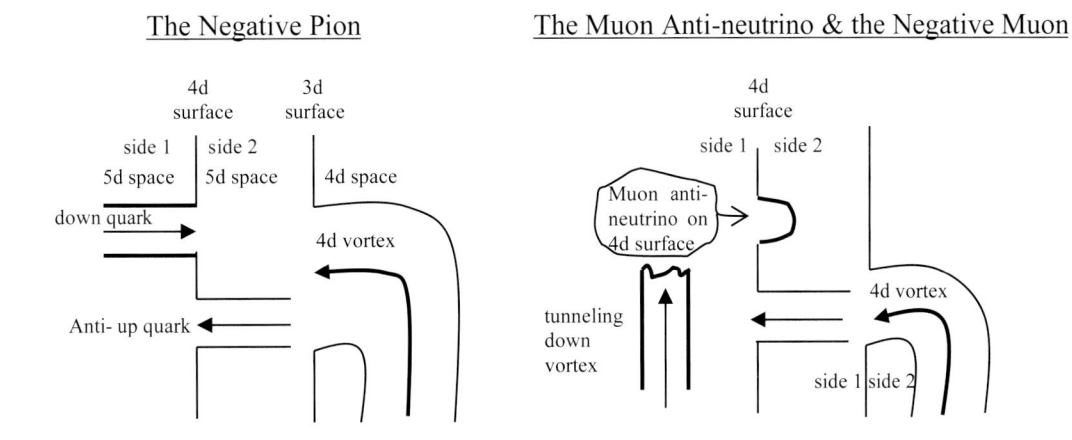

When the down quark deflates, its outward flowing 5d space into side 2 [note the direction of the arrow in the down quark] pushes side 2 inward. This inward push creates a quantized volume of 5d space bent into side 2 that becomes the muon anti-neutrino. [Note how the muon anti-neutrino is created upon the surface of 4d space.] The removal of the 5d volume of space causes its 4d surface [the 4d volume within the pion] to reduce in size, forcing the 4d vortex to reduce in size becoming the negative muon.

This reaction is represented as **equation 2**: π- → $\overline{V}_u$ + μ-

Note that the decaying down quark pushed the surface of 4d space outward towards side 2 in an opposite direction to that of the space flowing in the 4d vortex of the negative pion. This bend in 4d space *towards side 2* [previously designated -1] gives the muon anti-neutrino a lepton # of -1; and the direction of the flowing space in the negative muon vortex *towards side 1* [previously designated +1] gives the positive muon a lepton # of +1. Because the positive pion has a lepton # of 0, when all three are plugged into equation 2: **0 = -1+1** Hence, the lepton number is conserved.

We have also seen the first reason why a muon anti-neutrino is created during the creation of a negative muon.

Note: the down quark is tunneling with the anti-neutrino. Note too, the Vortex Theory reveals that the negative muon is a combination an anti-up quark and the extra dense region of space needed to change its -2/3 charge into a -1 charge.

Next, the reason why the muon anti-neutrino and the electron neutrino are both created are examined during the decay of the positive muon,. These relationships are important because they reveal how the space flowing into a decaying up quark vortex pulls the opposite side of space outward, creating an anti-neutrino. It also reveals how a denser region of space pulled towards a flowing vortex rebounds, pulling the opposite side of space inward, creating an electron neutrino. These relationships are examples of how rules #7 and # 3 from TABLE 52-1 are created:

THE EXPLANATION OF RULES #7 & #3 FROM TABLE 52-1: *why positive muons are accompanied by neutrinos; positrons are accompanied by neutrinos:*

Figure 52.4 [PART I OF THE DECAY]: <u>In the first part of this decay, the deflation of the up quark in the positive muon creates a muon anti-neutrino:</u> $\overline{V}_u$

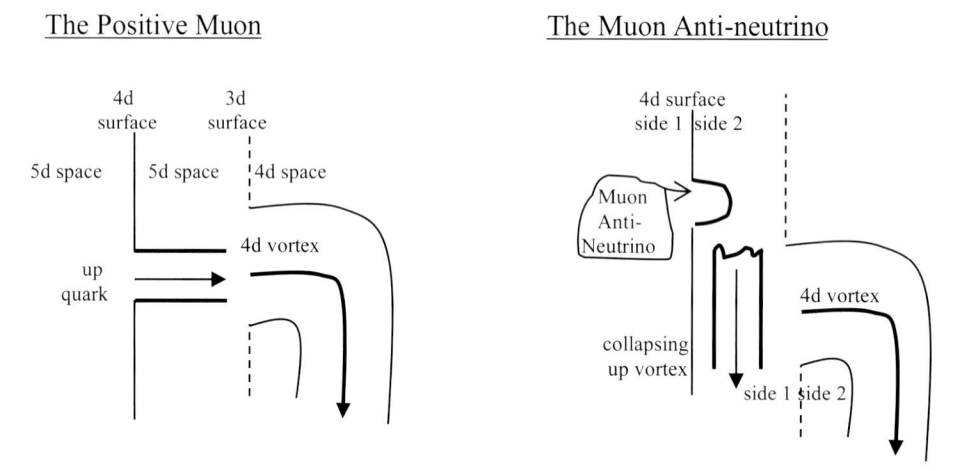

The Positive Muon   The Muon Anti-neutrino

When the up quark deflates, its inward flow - into side 2 - pulls side 2 inward. This inward pull creates a volume of space bent into side 2 that becomes the *muon anti-neutrino. However, this is only the first half of the decay.*

The second half of the decay is created by the deflation of the extra volume of space pulled into the volume of the 4d vortex needed to increase the density of the space in the vortex. This extra volume of 4d space within the positive muon was needed to change the charge of the up quark from a value of +2/3 to a value of +1. However, unlike the decays of the quarks that all bend space in the direction of the space flowing in their vortices; the decay of this extra volume of the 4d space bends the 3d surface of space outward in the *opposite direction* to the flow of space in the vortex.

This reversal occurs because unlike the collapse of the quark 5d vortices, the 4d vortex does not collapse completely. [If it had collapsed completely, the resultant *bend in the 3d surface* would be a single combination of the subtraction of the two opposite bends of the 3d component of the muon anti-neutrino and the 3d region of bent space created by the electron neutrino Consequently, the extra volume of 4d space that was just about to flow into the collapsed section of the vortex suddenly reverses its flow. This reversal pulls backward upon the 3d surface distorting it outwards towards side 1.This outward distortion creates a volume of space bent into side 1 that becomes the *electron neutrino; the shrunken vortex becomes the positron*:

Figure 52.5 [PART 2 OF THE DECAY]: <u>creation of the electron neutrino and the positron</u>

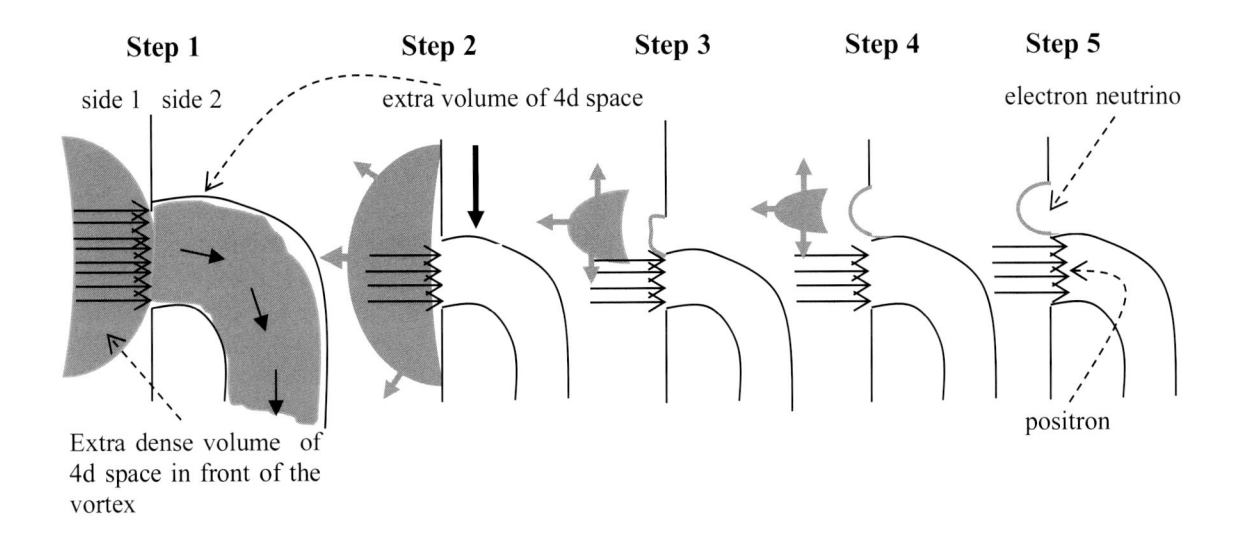

As seen in Figure 52.5, the extra volume of dense 4d space being pulled into the vortex to change the value of the up quark's +2/3 charge to the +1 charge [crescent shaped shaded area in step 1] of the positive muon reverses direction [step 2] when the vortex collapses. It now pushes back outward into the 4d volume [step 3]; this sudden outward push from side 2 towards side 1, pulls side 1 outward in the opposite direction. Again, this outward pull creates a volume of space bent into side 1 that becomes the electron neutrino, and the shrunken vortex becomes the positron.

Figure 52.6 <u>The simplified version of the creation of the electron neutrino: V_e</u>.

To simplify the sequence of events pictured in Figures 52.2 & 52.3, Figure 52.6 is now used. The gray volume of space in front of the vortex in [A] represents the extra dense volume of space that reverses direction and pushes backward into the 4d volume of side 1 [B], creating the electron neutrino [C].

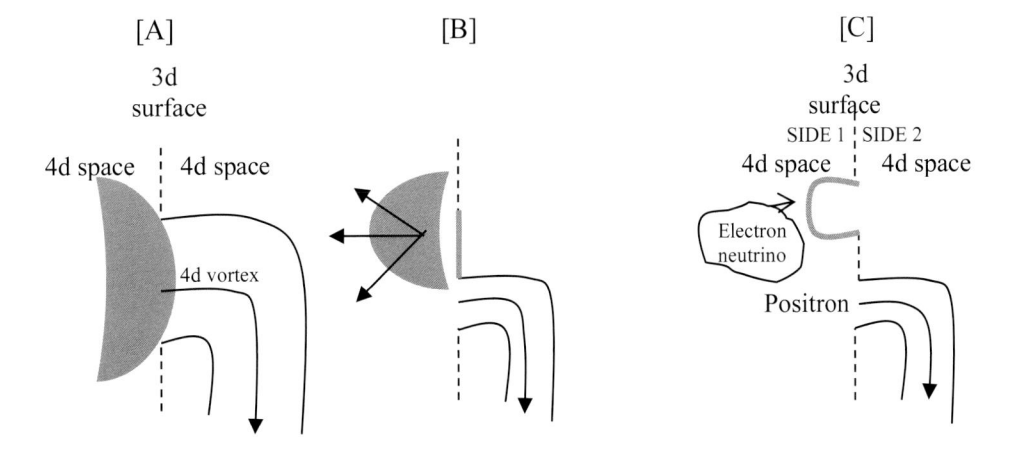

This reaction is represented as **equation 3**: $\mu^+ \rightarrow \overline{V_u} + V_e + e^+$

164

CREATION OF CONSERVATION OF LEPTON NUMBER:

In Figure 52.4, note how the direction of the space flowing in the decaying up quark's vortex pulled the surface of 4d space inward towards side 2. Note too that this direction is the *same* as the direction of the space flowing in the 4d vortex of the positive muon [towards side 2]. This bend in 4d space towards side 2 [previously designated -1] gives the muon anti-neutrino a lepton # of -1; and the direction of the flowing space in the positive muon vortex towards side 2 [previously designated -1] also gives the positive muon a lepton # of -1. [The creation of a muon anti-neutrino with a positive muon upon opposite sides of the reaction is an example of rule #7 in TABLE 52-1.]

Next, in Figure 52.5 [the second half of the decay], notice how the extra volume of space located just outside of the positive muon, [the extra volume needed to change the charge of the up quark from +2/3 to a value of +1], pushes backward into the 4d space from which it was originally pulled. This backward push pulls the 3d surface towards side 1 creating the electron neutrino. Observe too, how this outward push is in the opposite direction to the space flowing in the positive muon's vortex. Because this same vortex sheathing the positive muon deflates to become the positron's vortex, the direction of the flow of the positron's vortex is oriented in a direction opposite to the direction that the electron neutrino is oriented. This decay gives us our first insight into why an electron neutrino is paired with a positron when both are on the same side of the equation [affirming rule #3 in TABLE 52-1].

[Note too, in contrast to the muon anti-neutrino that is formed upon the 4d surface and bent into the 5d volume of space, the electron neutrino was created upon the surface of 3d space and is bent into the 4d volume.]

And finally, without having to look at our table of lepton numbers, but instead, just by looking at the direction the neutrinos are bent into the two sides of space and the direction space is flowing within the lepton vortices, we come up with the following lepton numbers: positive muon = -1; muon anti-neutrino = -1; electron neutrino = +1; and the positron = -1. When all four are substituted into equation 3: $\mu^+ \rightarrow \overline{V}_u + V_e + e^+$ the results are: -1 = -1 +1 -1 And we find that we have come up with the exact same result had we used our table.

Next, we shall see how a muon neutrino and an electron anti-neutrino are created during the decay of a negative muon.

THE EXPLANATION OF HOW RULES #8 and #4 FROM TABLE 52-1: *negative muons are accompanied by neutrinos; electrons are accompanied by anti-neutrinos:*

Figure 52.7 PART 1: <u>decay of the anti-up quark in the negative muon creates a muon neutrino: V_u</u>

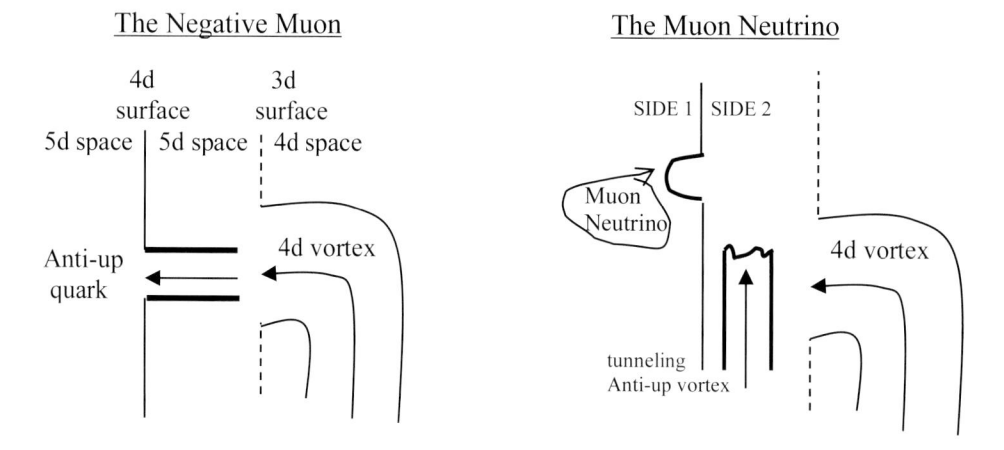

The decay of the anti-up is shown during the decay of a negative muon μ^-.

When the anti-up quark deflates, its outward flow – towards side 1 – pushes side 1 outward. This outward push creates a volume of space bent into side 1 that becomes the muon neutrino. However, this is only the first half of the decay.

The second half of the decay is created by the deflation of the extra volume of space in the volume of the 4d vortex needed to increase the density of the space in the vortex [see Figure 52.8]. This extra volume of 4d space within the negative muon was needed to change the charge of the anti-up quark from a value of -2/3 to a value of -1. However, unlike the decays of the up, down, and anti-down quarks that all bend space in the direction of the space flowing in their vortices; the decay of this extra volume of the 4d space bends the 3d surface of space outward in the *opposite direction* to the flowing space in the vortex.

This reversal occurs because unlike the collapse of the quark vortices, the 4d vortex does not collapse completely. [If it had collapsed completely, the resultant bend in the 3d surface would be a single resultant combination of the two opposite bends of the electron anti-neutrino and the 3d component of the muon neutrino.] Consequently, the extra volume of 4d space flowing out of the negative muon reverses its flow. This reversal pushes outward upon the 3d surface distorting it outwards towards side 2. This outward distortion creates a volume of space bent into side 2 that becomes the electron anti-neutrino; the shrunken vortex becomes the electron.

Figure 52.8 PART 2: creation of the electron and the electron anti-neutrino:

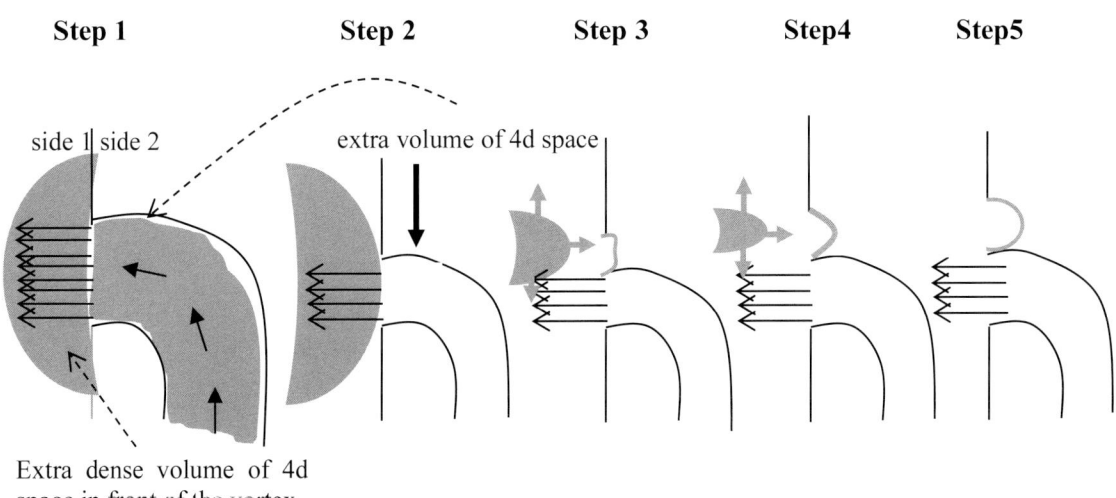

As seen in Figure 52.8, the extra volume of dense 4d space being pushed out of the vortex [crescent shaped shaded area in step 1] of the negative muon reverses direction [step 2] when the vortex collapses. It now pushes back outward into the 4d volume of side 2 [step 3]; this sudden outward push from side 1 towards side 2 pushes side 2 outward in the opposite direction. And again, this outward push into side 2 creates a volume of space bent into side 2 that becomes the *electron anti-neutrino, and the shrunken vortex becomes the electron.*

Figure 52.9 Creation of the Electron Anti-neutrino $\overline{V}_e$:

The creation of the electron anti-neutrino can be seen in the simplified drawing below:

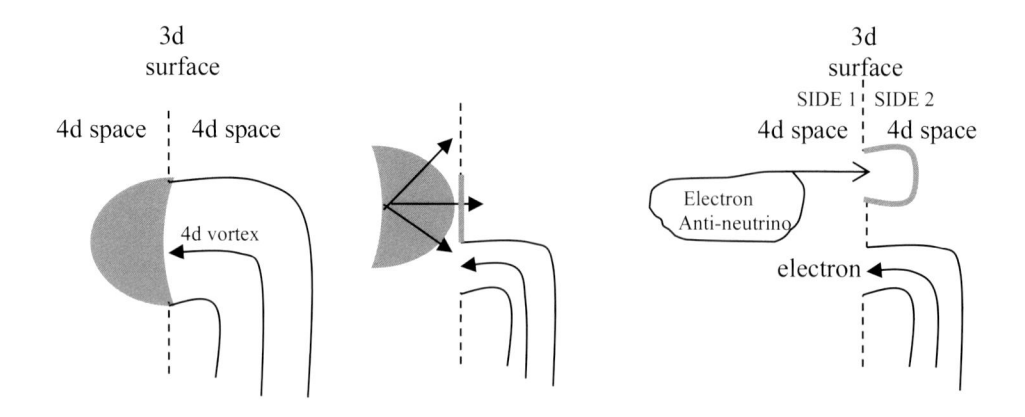

This reaction is represented as **equation 4**: $\mu^- \rightarrow V_u + \overline{V}_e + e^-$

CREATION OF CONSERVATION OF LEPTON NUMBER:

In Figure 52.7, note how the direction of the space flowing in the decaying anti-up quark's vortex pushed the surface of 4d space outward towards side 1. Note too that this direction is the *same* as the direction of the space flowing in the 4d vortex of the negative muon [towards side 1]. This bend in 4d space towards side 1 [previously designated +1] gives the muon neutrino a lepton # of +1; and the direction of the flowing space in the negative muon vortex towards side 1 [previously designated +1] gives the negative muon a lepton # of +1. This decay also reaffirms Rule #8 in Table 52-1.

Next, in Figure 52.8, [the second half of the decay], notice how the extra volume of space pushing outward into the 4d volume from the negative muon [that was needed to change the -2/3 charge of the anti-up quark to a value of -1] reverses direction and pushes outward into side 2 creating the electron anti-neutrino. Observe how the anti-neutrino's outward bulge is in the opposite direction to the space flowing in the negative muon's vortex. Because this same vortex sheathing the negative muon deflates to become the electron's vortex, the direction of the flow in the electron's vortex is oriented in a direction opposite to the direction that the electron anti-neutrino is oriented. Not only can we now understand, but we can also *see* why an electron anti-neutrino is paired with an electron when both are on the same side of the equation. This also gives an explanation for rule #4 in Table 52-1. [Note too, in contrast to the muon neutrino that is formed upon the 4d surface and bent into the 5d volume of space, the electron anti-neutrino was created upon the surface of 3d space and is bent into the 4d volume.]

And finally, without looking at our table of lepton numbers, but instead, just by looking at the way the leptons are bent into the two sides of space we come up with the following lepton numbers: negative muon = +1; muon neutrino = +1; electron anti-neutrino = -1; and the electron = +1. When all four are substituted into equation 4, $\mu^- \rightarrow V_u + \overline{V}_e + e^-$ and the results are: +1 = +1 -1 +1 And we find that we have come up with the exact same result we would have come up with had we used our table.

Next, one of the most famous reactions in particle science is the collision between a proton and an anti-neutrino. When the proton and anti-neutrino collide, they produce a neutron and a positron. This collision can now be explained using our understanding of the two sides of space and the way anti-neutrinos are created. The collision also provides an excellent example for the explanation of rule #5 in TABLE 52-1.

THE EXPLANATION OF RULE #5: *positrons are accompanied by anti-neutrinos*

Figure 52.10 The collision between a proton and an anti-neutrino:

In the drawings below, the anti-neutrino approaches and then strikes the proton:

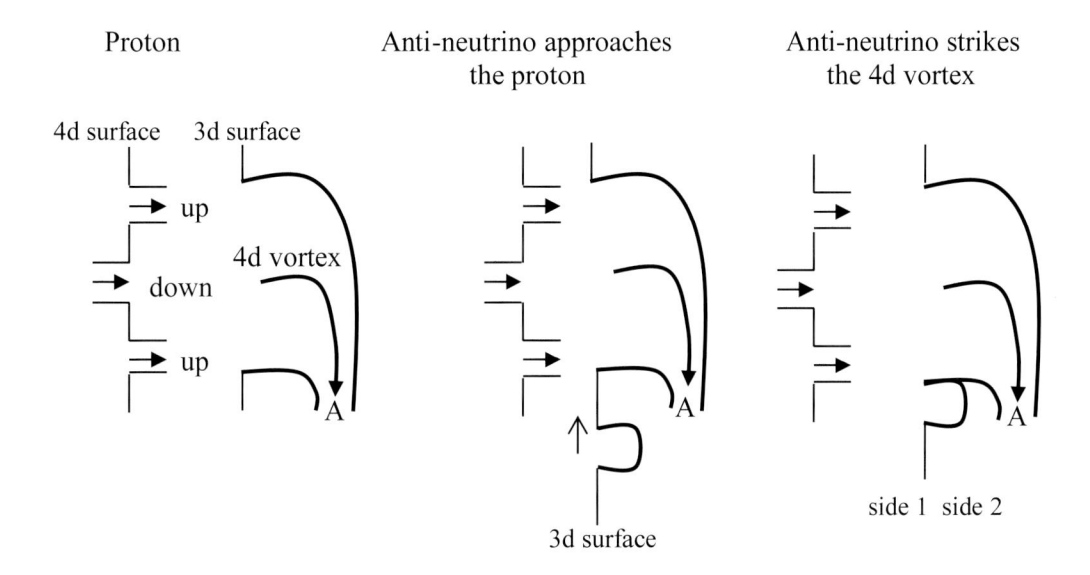

Moving at the speed of light, when the anti-neutrino strikes the 4d vortex, *its volume of space bent into side 2 is suddenly added* to the 4d vortex's volume of flowing space that is also bent into side 2. This almost instantaneous inflation causes three simultaneous events to occur:

#1. The 4d vortex breaks. One end curls back upon itself creating a 4d torus; while the other end continues to flow into side 2 becoming the positron:

Figure 52.11

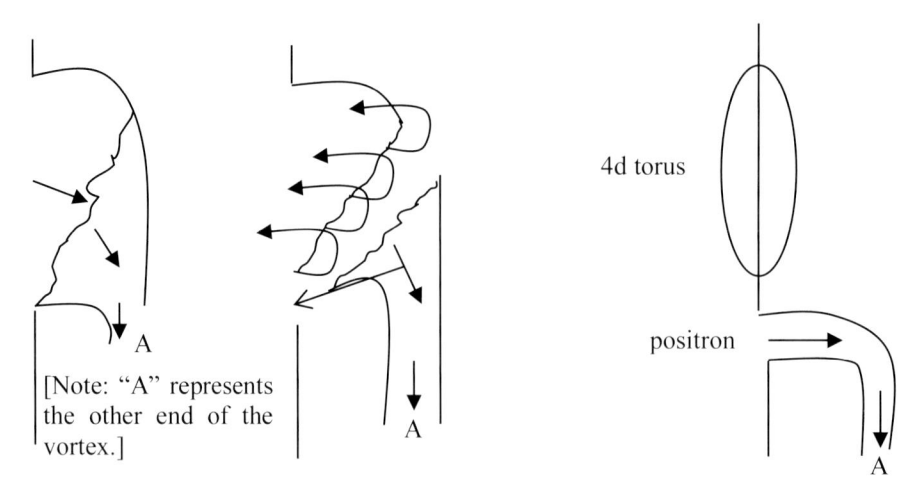

[Note: "A" represents the other end of the vortex.]

4d torus

positron

#2. As the electron anti- neutrino collides with the proton, its volume of 4d space bent into side 2 suddenly adds to the volume of 4d space that is bent into side two. This almost instantaneous increase in the volume of the space within the proton creates the effect of a down quark [a down quark adds 4d space to the 4d volume]. This causes the up quark nearest to the collision to immediately decay. The decay pulls side 1 outward creating the down, anti-down combination.

As the down, anti-down quarks are being created, and the breaking 4d vortex forms a torus, a circulating flow within the newly forming neutron begins; also, the anti-down and the decaying up quark tunnel off into space together - side by side away from the site of the collision.

[Note: it appears as if the anti-down quark and the up quark should unite to form a positive pion inside the vortex created by the positron. However, there is not a large enough volume of 4d space present. If there were, the increased volume would allow the positron's vortex to increase in size, allowing the anti-down and the up quark to remain within it creating a positive pion. However, this effect is indeed created during the decay of nucleons where N→Nπ

#3. When the up quark deflated, creating the down quark, the volume of space within the neutron is suddenly filled up. Consequently, the outward flowing space at the other end of the down vortex, [the anti-down quark], has no place to go [from the explanation of the Conservation of Baryon Law: See Chapter 53]. It cannot stay within the neutron, so upon its formation, the anti-down quark immediately deflates and tunnels through 4d space. As it deflates, its inward flowing space causes side 1 to pull inward, creating the <u>effect</u> of a muon neutrino. However, because the outward flowing space of the deflating up quark pulls side 2 inward, creating the <u>effect</u> of a muon anti-neutrino, as seen from the 3d surface, *the two opposite effects cancel each other and as the two quarks tunnel off together, no neutrino is seen, all that is left is the neutron and the positron:*

Figure 52.12 **Figure 52.13**

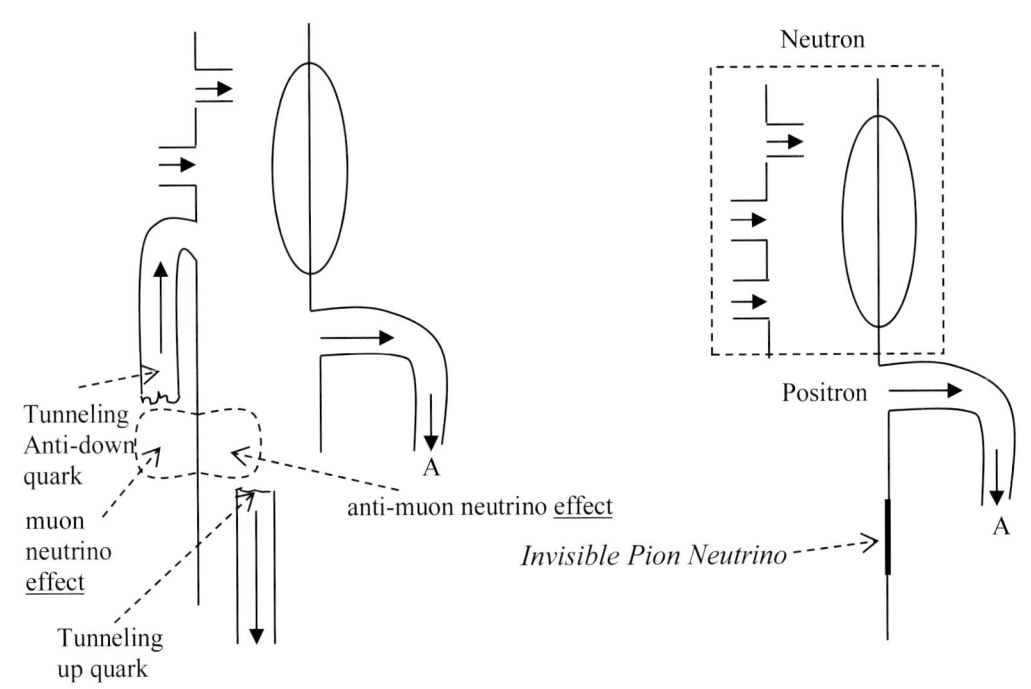

[Note: because the tunneling anti-down quark and the tunneling up quark travel side by side, their simultaneous opposite pulls upon the opposite surfaces of space cancel, and no neutrino or anti-neutrino is created. They are invisible. <u>But if they collide with a positron, using the principles of the Vortex Theory, it is theorized that they will create a positive pion</u>. **BECAUSE THIS TUNNELING DUAL ANTI-DOWN & UP QUARK REPRESENTS A NEW PARTICLE IN NATURE, IT WILL BE THE SUBJECT OF FURTHER EXPLORATION]**

CREATION OF CONSERVATION OF LEPTON NUMBER:

The collision between the proton, the electron anti-neutrino, and the resultant particles can be expressed in **equation 5**: $P^+ + \overline{V_e} \rightarrow N + e^+$.

Looking at Figure 52.10, it can be seen that the electron anti-neutrino is bent outward towards side 2, giving it a lepton number of -1; the proton is a baryon and has a lepton number of 0; the neutron is also a baryon possessing a baryon number of 0; and by observing the flow of space in the positron towards side 2, we can see that it has a lepton number of -1. When all four values are put into equation 5, the results are: $0 + -1 = 0 + -1$. And we find that we have come up with the exact same result had we used our table.

A PROTON, <u>MUON</u> ANTI-NEUTRINO COLLISION

[When the muon anti-neutrino containing a <u>tunneling up quark</u> strikes the proton, a similar set of circumstances is created. However, because the up quark in the muon anti-neutrino contains a component of 4d space, the 4d volume within the proton is further increased, increasing the size of the positron at the other end; allowing this newly arrived up quark to stay within the positron – forming the positive muon.

When the muon anti-neutrino containing a <u>tunneling down quark</u> strikes the proton, the down quark suddenly replaces the up quark, the vortex begins to break forming the neutron; and the additional 4d volume of space provided by the down quark's 4d component, adds to the forming positron's 4d volume. This additional volume allows the displaced up quark room within the forming positron to create a positive muon instead of a positron.]

THE EXPLANATION OF RULE #6: *electrons are accompanied by neutrinos:*

In this last section, we will explain rule #6 using the collision between the neutron and the electron neutrino; and in addition, explain the creation of the anti-neutrino during neutron decay. The collision between the neutron and the electron neutrino is discussed first.

THE COLLISION BETWEEN A NEUTRON AND AN ELECTRON NEUTRINO

When an electron neutrino collides with a neutron, the reverse set of events to what we have just seen occurs:

Just as the electron anti-neutrino increased the volume of the proton creating the chain of events that eventually turned it into the neutron, the electron neutrino decreases the volume within the neutron reversing the chain of events; allowing the neutron to turn back into a proton:

Figure 52.14 The electron neutrino strikes the neutron:

As the electron neutrino collides with the neutron, its volume of 4d space bent into side 1 suddenly subtracts from the volume of 4d space that is bent into side 2. This almost instantaneous decrease in the volume of the space within the neutron creates the effect of an up quark [up quarks remove 4d space from the 4d volume]. This causes the down quark nearest to the collision to instantly decay. The decay pushes side 2 outward creating the up, anti-up combination.

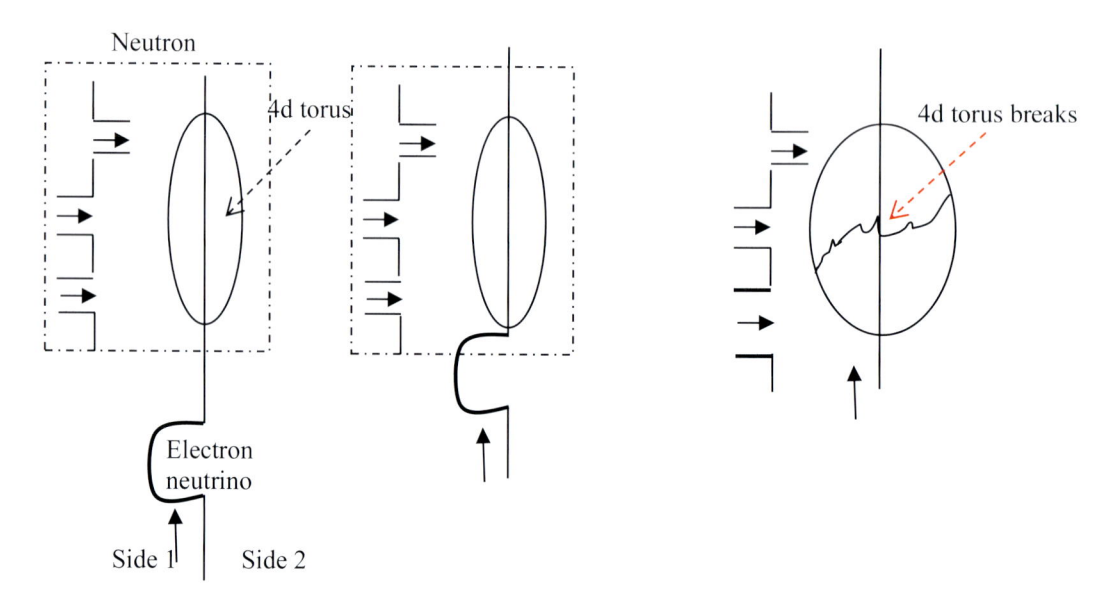

Figure 52.15 As the torus breaks…

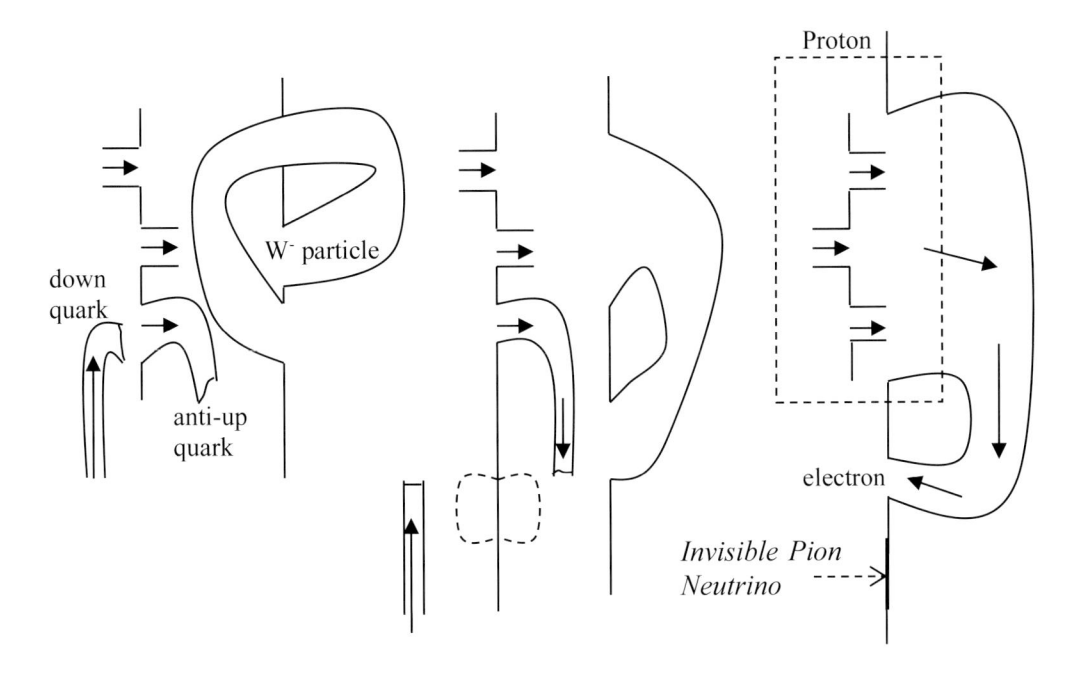

As the torus breaks and begins to reform the two ends of the 4d vortex, because the outward flowing space at the other end of the up vortex, [the anti-up quark], adds to the volume of the proton, it cannot stay within the reforming proton. Instead, it immediately deflates and tunnels through 4d space. As it deflates, its outward flowing space causes side 1 to push outward, creating the effect of a muon neutrino. However, and at the exact same time, because the down quark has also deflated, its outward push into side 2 creates the effect of a muon anti-neutrino. These two opposite effects cancel each other, and no neutrino is seen. All that is left is the proton and the electron seen below:

[*Note: because the tunneling anti-up quark and the tunneling down quark travel side by side, their simultaneous opposite pulls upon the opposite surfaces of space cancel, and no neutrino or anti-neutrino is created. They are invisible. But if they collide with an electron, using the principles of the Vortex Theory, it is theorized that they will create a negative pion.* **BECAUSE THIS TUNNELING DUAL ANTI-UP & DOWN QUARK ALSO REPRESENT A NEW PARTICLE IN NATURE, IT WILL BE THE SUBJECT OF FURTHER EXPLORATION**]

[Note, the breaking 4d torus creates the W $^-$ Particle, and W particles are collapsing 4d toruses]

[*Because the tunneling down quark and the tunneling anti-up quark travel side by side this new particle in nature is tentatively called an invisible Pion Neutrino.*]

CREATION OF CONSERVATION OF LEPTON NUMBER:

The collision between the proton, the electron anti-neutrino, and the resultant particles can be expressed in **equation 6**: $N + \overline{v}_e \rightarrow p^+ + e^-$

Looking at Figure 52.13, it can be seen that the electron neutrino is bent outward towards side 1, giving it a lepton number of +1; the proton is a baryon and has a lepton number of 0; the neutron is also a baryon possessing a baryon number of 0; and by observing the flow of space in the electron towards side 1, we can see that it has a lepton number of +1. When all four values are put into equation 6, the results are: 0 + 1 = 0 + 1. And we find that we have come up with the exact same result we would have had we used our table.

NEUTRON, MUON NEUTRINO COLLISION

When the muon neutrino containing a tunneling anti-up quark strikes the neutron, the following simultaneous events occur: the vortex breaks, the down decays creating the up, anti-up pair; the tunneling anti-up from the neutrino joins the breaking end of the vortex that would have become an electron, creating a negative muon instead; while the up joins the other up and down to form the proton as the decaying down and the newly formed anti-up tunnel off together to form an invisible pion neutrino. [Note: the tunneling anti-up does not annihilate with the up quark in the neutron because the position of the down quarks shield it.

When the neutron is struck by a muon neutrino containing a tunneling anti-down quark, a completely different set of circumstances is created. The anti-down quark and one of the neutron's down quarks annihilate forcing side 2 inward - creating an up, anti-up pair of quarks. The up quark combines with the other up quark and the other down quark to form a proton, while the anti-up quark combines with the forming electron to produce the negative muon. Since no other quarks are left, no invisible pion is created.

Chapter 53
The Explanation of the Law of Conservation of Baryons

> The law of the conservation of baryons states that the number of baryons must always remain constant. To express this statement mathematically baryons are given values of +1, their anti-particles given values of -1, and all other particles are given values 0. When expressed as an equation, the values of all the particles involved in the reaction must equal the values of all the particles left after the reaction is over.

The law of the conservation of Baryons can be traced to the longevity of the proton, and the discovery that more massive quarks are sheathed within up and down quarks.

It is postulated that the half-life of the proton is approximately 10^{31} years! I use the word postulated because nobody will be around long enough to observe if it is true or not. But no matter if it is right or wrong, what is important is the fact that a baryon with three quarks does not want to decay into other particles containing less than two quarks, but only ones containing three quarks. So why is this?

The answer could not be given until the discovery of the Vortex Theory. The Vortex Theory reveals that more massive quarks such as strange, charm, top, and bottom quarks are really higher dimensional holes within the lower dimensional holes called up and down quarks. Consequently, even though a Baryon might be listed as having an up, down and charm quark; in reality what it really has is an up, down, and up – containing a charm quark. So when the charm decays, it decays into the up quark it was contained within. Turning it back into a proton: which it always was!

The same is true for the neutron. Although the neutron has a down, up, and another down quark, one of these quarks might possess another higher dimensional hole within it: such as a strange, charm, top, or bottom quark. However, when the higher dimensional hole decays, it will leave the original three quarks: down, up, and down. That will eventually decay into a proton, electron, and anti-neutrino.

Consequently, it can be seen that any baryon is ultimately a proton or neutron in disguise, and will decay back into a proton or neutron, with the free neutron ultimately decaying into a proton.

During particle collisions, the stable relationship between the up, down and up quark, keeps it from breaking apart. The down between the two ups, shields the positive charges of the two ups from each other, and keeps them from flying apart. This stable configuration of the three quarks is what gives it longevity.

Figure 53.1

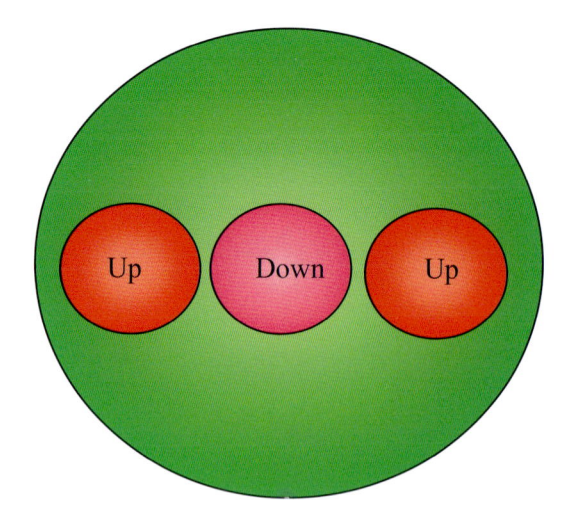

Chapter 54
The Explanation of the Conservation of Momentum

> Unbeknownst to most viewers, the conservation of momentum is seen every day in every sporting event on television. When the man kicks a football or hits a baseball with a bat, it is not the man, but the law of the conservation of momentum that decides the outcome of the play! However, none of the athletes in the game, or the professors in the universities where they play, knew how angular momentum was created until now!

It was decided to explain the conservation of momentum after the explanation of Newton's three laws and the Conservation of Angular Momentum to simplify its explanation. However, because hindsight is always 20/20, this order of explanation now seems unimportant. Because using the principles of the Vortex Theory, the explanation of the conservation of momentum is now so simple to explain, it almost seems trivial!

The transfer of part or all of the internal waves within the surfaces of the three and fourth dimensional holes of one object to another object, is the explanation of the Conservation of Momentum.

The reason why the momentum formula ($P = MV$) works is a result of the "number of holes" present and their velocity [note, mass is only proportional to the number of holes].

According to conventional wisdom, when a large mass (m) is moving at a certain velocity (v), it possesses a certain amount of momentum (P). If the mass increases the momentum increases, and if the velocity increases the momentum also increases.

But when seen by the Vortex Theory, there is another explanation for this equation.

Because a mass is roughly proportional to the number of holes an object possesses, the larger the number of holes, the more waves there are, and the larger the amplitude of the collective wave: increases (P).

When the same number of holes move forward at a faster velocity, the larger the amplitude of the individual waves, the larger their addition, and the larger the value of the amplitude of the collective wave: increasing (P).

When one object strikes another, part or all of the waves are transferred from one object to another. It is this transfer that creates the Conservation of Momentum.

When two objects collide, the transference of some of the waves is reflected back to the holes of the original moving object, and the direction of the original object is changed.

It is the wave within the surface of each individual hole the atom is made out of that is the real explanation of Conservation of Momentum. The amount of momentum transferred to another object is a function of the percentage of the collective sum of these individual waves that are transmitted to other objects during collisions.

Chapter 55
The Explanation of the Conservation of Angular Momentum

> One of the world's most enduring mysteries is finally explained.

The explanation for the conservation of angular momentum has been waiting for a long time. Although our ancient ancestors did not know its technical name, they were able to observe its effects. Every time a person laid down upon the ice of a frozen lake, grasped the outstretched arm of a friend, allowed themselves to be spun around and then quickly pulled in their arms and legs, the conservation of angular momentum went to work and they began to spin rapidly. This pirouette motion is most spectacular when performed by a highly skilled ice skater.

Although this fundamental law of nature could be expressed with mathematical equations, there was still no explanation of how it worked until now. And it is a most fascinating explanation:

When a three dimensional hole such as a proton moves through space, the internal wave upon its three dimensional surface [and the fourth dimensional waves within the fourth dimensional holes] moves it in a straight line.

However, when this distorted hole is part of say a ball attached to a string held in the hand of a man who is whirling it around his head, the hole is now moving in a circular path. The wave within the hole wants to continue to move in a straight line, and if the string broke, the hole would again continue to move in a straight line starting from the point of the break. However, because the string is strong, the ball moves at a constant speed, and the wave is reconfiguring at a constant rate. But if the man (while continuing to whirl the ball on the string around his head with his right hand) holds the other end of the string in his left hand and slowly starts to pull upon it, the length of the string between the man's right hand and the ball begins to shorten. This shortening of the string shortens the balls radius' of rotation.

As the string pulls upon the ball and the radius of rotation gets shorter, the pull on the string creates an acceleration that is transferred to all of the holes within the ball including the hole or proton we are observing. This acceleration creates an *additional distortion within the hole*, adding to the distortion that is already there, amplifying the size of the internal wave, making its space reconfigure faster, making it move faster.

The addition of these two distortions can be seen in the following drawings: Figure 55.1 shows the distortion if the hole was moving through space in a straight line; Figure 55.2 shows the added distortion created by the pull from the string; Figure 55.3 shows the vector addition of the two distortions; and Figure 55.4 shows the resultant vector distortion.

Figure 55.1

Figure 55.2

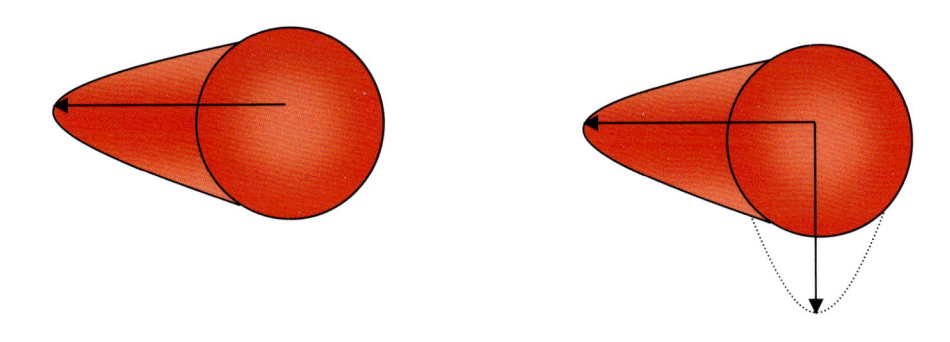

Figure 55.3

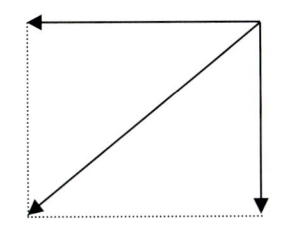

Figure 55.4

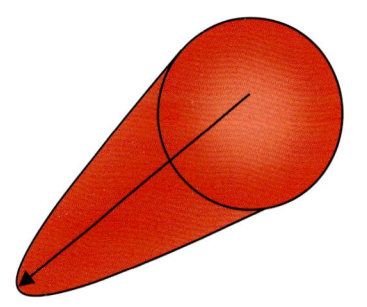

Because an acceleration has increased the amplitude of the internal wave within the three and fourth dimensional holes, the ball is now moving faster. As the string continues to shorten, the larger amplitude of the increasing wave continues to make it move even faster.

But so far we have only been describing what is happening within the ball.

As the man pulls upon the string and the string pulls upon the ball, the ball pulls back upon the man through the string. Because the pull of the ball also distorts the holes within the man, his motion is slightly accelerated too. However, because the ball possesses a small number of holes in relation to the holes within the man, the distortion it transfers to the holes within the man is a lot less than the distortion transferred to the ball through the man.

Nevertheless, the distortion has been made within the holes within the man, changing his rotational speed, slightly accelerating him too.

But when the reverse happens, the distortions in the holes in both the man and the ball are ameliorated.

When the reverse happens, and the man loosens his grip on the string, allowing the ball to move away from him, the distortions are reversed.

Assuming a perfect transference of distortions [where none is lost to friction and no energy is dissipated as heat], as the man begins to let the ball move away from him, the holes it is made out of begin to be slightly distorted outward, in the opposite direction from whence they were originally distorted. This outward distortion negates their original distortion, and the ball begins to move slower. It is the same for the man.

When he began to loosen his grip upon the string and allow the ball to move away from him, its pull upon him lessoned as well. This lessoning of the pull decreased the distortions created within his holes, decreased their acceleration, and slowed his rotational speed.

Although the above illustration dealt with an example of a ball whirled above the head of a man, it must be noted that the exact same situation will occur with the holes of two objects held together by gravity [such as the Earth and the Sun].

Consequently, it is now easy to see how the increase and decrease in the distortions in the holes of two different objects connected together – increase or decrease the amplitude of their internal waves, allowing their speeds to accelerate or decelerate – creating the law called the Conservation of Angular Momentum.

Chapter 56
The Explanation of the Conservation of Charge

> One of the great mysteries of the subatomic world is finally and easily explained.

The discovery of the two vortices of space flowing back and forth between the electron and the proton explains many observations that were unexplainable before. The most obvious of these is the "conservation of charge".

The term "conservation of charge" is based upon the scientific observation that the break-up of a particle, or the change in a particle from one type to another [such as the decay of the neutron into a proton, electron, and an anti-neutrino] always leaves the same net charge. For example, since the net charge upon the neutron was zero, when the neutron was hit by the neutrino, and changed into a proton and an electron, the respective charges of these two particles are positive and negative (the anti-neutrino has no charge). Hence, when a +1 and a –1 are added, they cancel each other out [+1 + (– 1) = 0]: making their net charge equal to zero.

For years, this *conservation of charge* was a mathematical observation in search of an explanation. But not any longer! The failure to explain this phenomenon resulted from a lack of knowledge about the existence of the vortex. But now, it can clearly be seen that these two "particles," (holes in space), are really the two ends of a swirling higher dimensional vortex constructed out of three dimensional space. Consequently, no matter how many times the vortex is broken, the number of additional entrances and exits into and out of three dimensional space will always cancel each other out, and their sum will equal the number originally present.

Hence it is easy to see how charge is conserved, and how another great mystery of science is easily explained.

Chapter 57
The Explanation of the Conservation of Mass and Energy

Perhaps the most famous equation in the world is $E = mc^2$. We found out how it works when we built the atomic bomb. But we did not know *why* it works until now.

The explanation of the conservation of mass and energy is one of the longest and most awaited explanations in all of science. Luckily, it is as easily explainable as the equation $E = mc^2$.

Mass is equal to the hole that is left when a volume of three dimensional space is removed to create the three dimensional hole; and *energy* is equal to the volume of the removed space.

Or, using ice cream as an example, taking a scoop out of the surface of a tub of ice cream leaves a hole in the surface: this hole is the equivalent of mass; while the volume of ice cream in the scoop is the equivalent of energy.

The expansion and contraction of the energy is explained by Planck's constant:

In the formula $E = hv$, h = Planck's constant, E = energy, and v = frequency of light. This formula can also be expressed as $h = E / v$. When the formula is changed, notice how Planck's constant is directly proportional to energy and inversely proportional to frequency. Since *h* is a constant, if the frequency of a photon increases, its "energy" [its dense volume of space] has to increase; and if the energy decreases, the frequency [ability to expand and contract] has to decrease.

Because the Vortex Theory reveals that energy, or rather photons are nothing more than "bunches" of very dense space surrounded by a massive region of dense space, Planck's constant is revealed to be the ratio between the amount of dense space that we call energy and its frequency of vibration [ability to expand and contract]!

PART V
QUANTUM ENTANGLEMENT

Chapter 58
The Explanation of Quantum Entanglement

> This great mystery is easily explained by tiny vortices existing between sub-atomic "particles."

As explained in the PhD Thesis in Book 1, what we call "particles" such as protons, anti-protons, electrons and positrons, are really three dimensional (3D) holes at the ends of tiny fourth dimensional (4D) rotating vortices we cannot see. Because of this relationship, each end rotates in the opposite direction to each other.

"Note: this effect is demonstrated by a 2 foot piece of PVC pipe rolling across a table. From one side of the table, it appears to be rotating clockwise; while from the other side, it appears to be rotating anti-clockwise."

REVERSAL OF SPIN:

Because the PVC pipe is made out of billions of atoms, like the links of a chain, each atom is twisted by the atom before it. Hence, a reversal in the direction of the rotation of the pipe is created out of the reversal spins of billions of individual atoms. However, the fourth dimensional vortex is one single structure. Consequently, its long length seems to reverse instantly. Its reversal also causes the other end to reverse its spin too.

QUANTUM ENTANGLED ELECTRONS:

When two electrons in the valence shell of an atom touch each other, their opposite rotations in three dimensional (3D) space begin to rotate between them. When they are ejected out of the atom, the rotational 3D space between them creates a 3D vortex. Just as with the 4D vortex, no matter what its length, remains one single structure made out of the same 3D space photons are made out of.

When one end (one electron) is hit by a photon, the length of the vortex suddenly tries to lengthen. However, the mass of the electron at the other end resists its movement: resisting the lengthening of the vortex. Subsequently, the second electron emits the same amount of volume equal to the 3D volume of space the original photon contained. Making it seem as if the original photon passed instantly between the two distant electrons faster than the speed of light.

PHOTON ENTAGLEMENT:

Similar to the 3D vortices created by the two electrons, 3D vortices are created when two photons of opposite spins collide. This does not happen often because photons are dense regions of space.

This creates anti-gravity effects, causing them to avoid each other. The same effects created by these vortices are similar to those described by electron vortices = twisting one end twists the other etc.

THE END

[Book 4 is next: The End of the Concept of Time PART 4: The Creation and Destruction of the Universe!]